Studies in Canadian Geography

Etudes sur la géographie du Canada

Ontario

Ontario

Edited by /Sous
la direction de
Louis Gentilcore

published for the 22nd International Geographical Congress
publié à l'occasion du 22e Congrès international de géographie
Montréal 1972

University of Toronto Press

© University of Toronto Press 1972
Toronto and Buffalo

ISBN 0-8020-1919-6 (Cloth)
ISBN 0-8020-6160-5 (Paper)
Microfiche ISBN 0-8020-0257-9

Printed in Canada

Contents

Foreword

The publication of the series 'Studies in Canadian Geography' by the organizers of the 22nd International Geographical Congress introduces to the international community of geographers a new perspective of the regional entities which form this vast country. These studies should contribute to a better understanding among scholars, students, and the people of Canada of the geography of their land.

Geographical works embracing the whole of Canada, few in number until recently, have become more numerous during the last few years. This series is original in its purpose of re-evaluating the regional geography of Canada. In the hope of discovering the dynamic trends and the processes responsible for them, the editors and authors of these volumes have sought to interpret the main characteristics and unique attributes of the various regions rather than follow a strictly inventorial approach.

It is a pleasant duty for me to thank all who have contributed to the preparation of the volume on Ontario. A special thanks is due to: Mr R.I.K. Davidson of the University of Toronto Press; Mr Geoffrey Lester, who guided the Cartography Laboratory of the Department of Geography, University of Alberta in preparing all the illustrations; the Canadian Association of Geographers for its financial support; and the Executive of the Organizing Committee of the 22nd International Geographical Congress. Finally I wish to thank Dr Louis Gentilcore, professor of geography at McMaster University, for having accepted the editorship of this volume.

LOUIS TROTIER
Chairman
Publications Committee

Avant-propos

Par la publication de cette série d'"Etudes sur la géographie du Canada,' les organisateurs du 22ᵉ Congrès international de géographie ont voulu profiter de l'occasion qui leur était donnée de présenter à la communauté internationale des géographes une perspective nouvelle des grands ensembles régionaux qui composent cet immense pays. Ils espèrent que ces études contribueront aussi à mieux faire comprendre la géographie de leur pays aux Canadiens eux-mêmes, scientifiques, étudiants ou autres.

Les travaux d'ensemble sur la géographie du Canada, peu nombreux jusqu'à récemment, se sont multipliés au cours des dernières années. L'originalité de cette série provient surtout d'un effort de renouvellement de la géographie régionale du Canada. Les rédacteurs et les auteurs de ces ouvrages ont cherché moins à inventorier leur région qu'à en interpréter les traits majeurs et les plus originaux, dans l'espoir de découvrir les tendances de leur évolution.

C'est pour moi un agréable devoir de remercier et de féliciter tous ceux qui ont contribué d'un manière ou d'une autre à la réalisation de cet ouvrage sur l'Ontario. Il convient de mentionner les membres du Comité d'organisation du 22ᵉ Congrès international de géographie; M. R.I.K. Davidson, des Presses de l'Université de Toronto; l'Association canadienne des géographes; le département de géographie de l'Université de l'Alberta, à Edmonton, dont le Laboratoire de cartographie a préparé toutes les illustrations de cet ouvrage sous la direction habile et dévouée de M. Geoffrey Lester. Je remercie enfin M. Louis Gentilcore, professeur de géographie à l'Université McMaster, d'avoir accepté d'assumer la direction de cet ouvrage.

LOUIS TROTIER
Président du
Comité des publications

Preface

Ontario is the most populous and most prosperous province in Canada. One-third of the nation's population lives here. They produce more than one-half of Canada's manufactured goods, one-quarter of her output from mines and forests, and one-third of the farm income. Accompanying this economic pre-eminence is a majestic primeval geography. Ontario extends through sixteen degrees of latitude and a distance of over 1600 kilometres, from barren tundra along a saltwater shoreline in the north to fertile lowlands bordering freshwater lakes in the south.

Productivity and size, two of the basic elements in the geography of the province, stand in contradiction to one another. The former is concentrated in a very small area with an identity and even a name of its own, 'Southern Ontario,' a portion of the province that is as overwhelming in its concentration of activity as the remainder is in its areal extent. The recognition of this distinction is a prerequisite to the further study of a subject which has been widely neglected, both in Ontario and in the rest of Canada. Writers and artists, historians and geographers have paid little attention to the province. It is a baffling region, one which 'has achieved a significant place in the Canadian sun, but no one quite knows what the place is, even though other areas would like to achieve the same position' (Warkentin 1966). The purpose of this short volume is to contribute to an understanding of Ontario, to point out something of what it is both to those who are already acquainted with the province and to those who are being introduced to it for the first time.

Seven Ontario geographers have addressed themselves to the task. Statements have been prepared on the province's physical environment, settlement, economy, cities, and political organization. Each of the chapters may be read as a separate unit. At the same time, some degree of integration has been achieved in the recognition of certain themes. One of these is the differentiation already noted. The contrast between a land of well-kept farms and growing cities in the south and one of meagre or no development beyond it follows from the descriptions of physical features and the course of growth presented in the first two chapters. The strength of the contrasts is confirmed and heightened in subsequent descriptions of the economy, cities, and political structure.

Operating within this first set of contrasts is a process of concentration, particularly active in Southern Ontario, but evident also in the north. Its performance begins with European settlement. As familiarity with the environment increases, population and agricultural production converge on better lands. Urban growth is highlighted by the emergence of a cluster of cities around the western end of Lake Ontario, commonly referred to as the 'Golden Horseshoe.' The ability of larger places to grow at the expense of smaller ones is an early and persistent part of geographic change. Its acceleration eventually produces the hungry cities of the twentieth century, growing roughly and rapidly, reaching into a variety of hinterlands, dominating them and then extending that domination. The climax is reached in the city of Toronto, focus of the Ontario heartland, and a metropolis both of the province and of the country. A description of the city is an indispensable and revealing part of any delineation of the province.

The increasing concentration of population into an urban system has posed special problems in the utilization of the Ontario environment. One aspect of these problems is the need to fit spatial units of government with those of economic activity. The discussion, as presented in this volume, has wider repercussions. It points to a fundamental consideration that emerges from a reading of all the chapters. Ontario society no longer perceives its environment solely in terms of livelihood. Habitability is increasingly valued. Hopefully, the chapters we have written not only will add to our knowledge of the Ontario environment but will also contribute to its more effective use.

The preparation of this volume and the research results summarized in its chapters have been supported by various sources: The Canada Council, The National Advisory Committee on Geographical Research, The Canadian Council on Urban and Regional Research, The Ontario Economic Council, The Environment Study (Bell Canada), Central Mortgage and Housing Corporation, and The Ontario Institute for Studies in Education.

This support is gratefully acknowledged.

McMaster University   R.L.G.
19 January 1972

1 The Environment

L.G. REEDS

The major contrasts in land use and settlement patterns in Ontario reflect its physical geography. Dominating the province areally is the Precambrian Shield, a region of shallow soils, extensive forests, sparse agricultural settlement, and dispersed urban centres. The Palaeozoic lowland of Southern Ontario stands out because of its more intensive agriculture, the greater density of transport facilities and the concentration of manufacturing and urbanization (Fig. 1.1). Despite advances in technology and management, primary resource development continues to be closely related to the physical environment. The purpose of this chapter is to interpret these relationships, to emphasize the necessity of conserving and using the natural resources wisely and of maintaining the good qualities of the environment.

Geology and Mining

The Precambrian Shield is a rugged upland composed of ancient igneous and metamorphic rocks which occupies a large part of Northern Ontario and which protrudes into Southern Ontario as far south as the St Lawrence River to the east of Kingston. The gentler lowland around Hudson and James Bay and in Southern Ontario is underlain by sedimentary rocks of younger Palaeozoic age.

The old crystalline rocks of the Shield were altered greatly during the past one billion years but now constitute a stable land mass. In many areas in Northern Ontario, volcanic and intrusive rocks occur and as Figure 1.1 reveals, these contain the chief mines and account for its emergence as one of the most productive mining areas of the world. Over 80 per cent of the total one billion dollar annual mineral production in Ontario comes from the metallic group in which nickel and copper are the most important. The mining industry in the Palaeozoic division is on a very small scale. On the other hand, the southern areas produce the bulk of the structural materials which are required in increasing quantities to provide for the needs of road construction and the building industry.

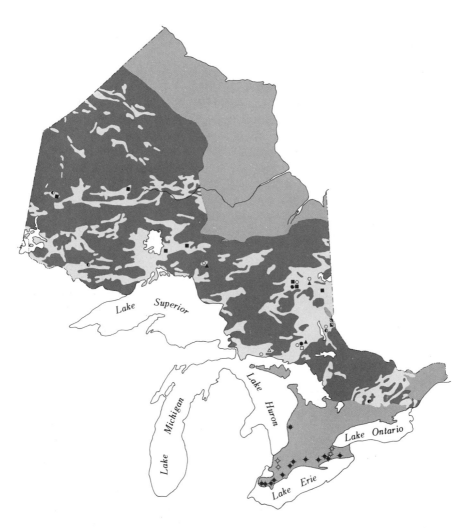

■	Precambrian - Mostly Granitic Rocks	

Nickel	●
Copper	○
Gold	■
Lead	□
Zinc	▲
Uranium	△
Iron Ore	◐
Silver	◐
Cobalt	◘
Salt	✳
Gypsum	✧
Gas	◆
Oil	◇

Precambrian - Mostly Granitic Rocks

Mesozoic and Palaeozoic Sedimentary Rocks

Mainly Volcanic and Sedimentary
(Most of the known mineral deposits
occur in this rock assemblage)

0	100	200	Miles
0	160	320	Kilometers

1.1
Geological Divisions and Chief Mines
(After the Ontario Department of Mines)

Landforms

The present landscape features of Ontario have been largely shaped by glaciation. Mammoth ice sheets of up to one mile or more in depth moved over the surface on several occasions and scraped off much of the unconsolidated material, particularly in the north, depositing it to form a great variety of landforms in the south. During the melting of the glaciers, several large lakes developed in which fine sediments were laid down to create the flattish plains or clay belts which occur in both northern and southern Ontario.

Northern Ontario lacks striking topographic forms and with the exception of a few rugged sections of the Shield, as in the vicinity of Lake Superior, the relief tends to be modest. Gradients of three to four feet per mile are common in the clay belts and in the Hudson Bay lowland. The latter is a low flattish, coastal plain underlain by horizontally bedded sedimentary strata covered by a variable depth of drift and dissected by shallow, meandering rivers. The main relief features are old beach ridges, and large areas are covered with poorly drained bog. The main forests are confined to the banks of the rivers and to the better drained ridges (Fig. 1.2).

The clay belts stand out because of their gentler relief, a general absence of rock outcrops and the presence of good forests and agricultural development. These, however, are not continuous clay areas but contain a variable amount of till and other materials. The largest area, the Great Clay Belt, of approximately 13 million acres centres on Cochrane while the Little Clay Belt of one million acres is located to the west of Lake Timiskaming. The Rainy River Clay Plain, the Thunder Bay Lowland, the Nipigon Basin, and the Nipissing Lowland also contain discontinuous areas of clay with agricultural settlement. The intervening uplands with a greater abundance of rock outcropping, shallow soils and variable sediments are mainly mining areas with some forestry and recreation.

The Shield of Southern Ontario, with its protruding rock knobs and intervening pockets of sand, silt, and clay, comprises about one-third of the total area of peninsular Ontario; it is primarily a recreational and forested area with no large urban places and widely dispersed settlement. The perception of the resource potentials of this region has changed with time. The advance of agricultural settlement into the Shield during the hectic lumbering era of the nineteenth century is described in chapter 2. Since the 1870s, there has been a continuing exodus of agricultural settlers. The region's recreational resources are now regarded as its most important asset.

The most striking topographic feature of Southern Ontario is the Niagara Escarpment which extends from Niagara Falls to the northern tip of the Bruce Peninsula and Manitoulin Island. Created by the differential erosion of the hard, Silurian dolomite and the softer, underlying shales, the Escarpment has provided a source of water power for early mills, hydroelectricity for modern factories, limestone for steel-making and for construction materials, a shelter for the Niagara fruit belt, and a site for the 240-mile Bruce Trail. The rocks of the Escarpment have provided the training ground for some of Canada's leading geologists. The study of its stratigraphy has thrown much light on the geology of the western parts of Southern Ontario and on the location of gas, oil, and non-ferrous mineral deposits. Fossils found throughout the various strata provide a record of the climate and life in prehistoric times. The valleys carved in its face by ancient rivers and their modification by glacial erosion provide some of the clues to an understanding of geological past in this part of Canada. Aware of the urgency to preserve this unique resource, the province is taking steps to ensure its maintenance and protection (Ontario Regional Development Branch 1968).

Most of the landforms of Southern Ontario other than the Niagara Escarpment are the result of glacial action (Chapman and Putnam 1966). These include the tumbled masses of stratified drift or kame moraine sorted by the gushing meltwaters of the wasting ice lobe and the mounds of till moraine pushed into position by the snowplow-like action of the advancing glacier. Represented as well are the fields of drumlins with their oval-shaped ridges of unsorted boulder clay that have been plastered into their characteristic inverted-spoon shapes under the active ice sheet. Beach ridges mark the former shorelines of higher-level glacial lakes. Many present day rivers and their tributaries occupy broad spillways that were carved by former immense meltwater streams.

Limestone plains from which much of the unconsolidated overburden was removed and where shallow and excessively stony soils have developed, flank the Shield in eastern and central Ontario and continue along the Niagara Escarpment in the Bruce Peninsula and on Manitoulin Island. Till plains with their drumlin fields and eskers and relief varying from gently undulating to strongly rolling occupy extensive areas in western and central Ontario. The morainic systems, including the Dummer Moraines of the Kawartha region, the Oak Ridges or Interlobate Moraine of the mid-section of central Ontario, and the Horseshoe Moraines of western Ontario have hummocky surfaces, uneven slopes, and materials ranging from stony tills to sands and gravels. Clay plains with flattish relief and generally heavy-textured soils occur in the Toronto region, in the Niagara Peninsula, and in southwestern Ontario. Sand plains are found along Lake

Erie, west of Lake Simcoe, in eastern Ontario, and in the Shield. Bogs and marshes occupy old lagoons along former shorelines, parts of spillways, or depressional areas in the moraine systems.

The Physical Environment and Agriculture
The distribution of landforms and their associated geological materials provide the key to an understanding of the present landscape. Soil drainage conditions and soil erosion are closely linked with surface configuration and materials. Type of farming, farm values, and productivity continue to reflect the influence of landforms, although socio-economic factors and technology have assumed greater significance.

The level to undulating land associated with the clay, till, sand, and limestone plains is conducive to intensive agricultural use provided soil qualities are favourable. Soil erosion is not generally a problem although drainage, shallowness or stoniness may be a serious limitation. On the other hand, the variable relief of the moraines and drumlins may prohibit intensive use for agriculture unless conservation practices are adopted. Matthews has appraised agricultural land-use hazards (Table 1.1) in Southern Ontario and has estimated the acreages which have significant limitations (Matthews 1956).

One should point out that in certain parts of Ontario, the natural drainage has been improved by artificial means and a highly productive agriculture has evolved as in the cash-crop areas of Essex and Kent counties. The poorly drained soils of eastern Ontario have not been improved as extensively and productivity is at a much lower level. Poorly drained soils often occur in association with better drained soils as in the case of inter-drumlin areas, spillways, marshes, and bogs. Many of these areas are left wooded or are used for permanent pasture. A good example of the effects of imperfect drainage on agricultural development can be cited in the mixed farming areas of the Niagara Peninsula, where the level of output per unit area has been affected by the problems related to the maintenance of a satisfactory tilth and to excess moisture in the soils at certain times of

Table 1.1 Agricultural land-use hazards (million acres)

Poor drainage	4
Imperfect drainage	4
Hilly relief	1
Low fertility and moisture	3
Susceptibility to water erosion	6
Excessive stoniness	1
Shallow over bedrock	2

the year. Areas which suffer from moisture deficiency and low inherent fertility include areas of sandy soils which require careful maintenance of organic matter content and adequate fertilization in order to be productive. The sand plain of Norfolk County and adjoining areas was a depressed area until the advent of tobacco-growing. Now it is one of the leading areas of Ontario from the standpoint of net dollar returns per acre of agricultural land. Excessively stony soils were used satisfactorily for a time in the days of horse-drawn machinery, but have been abandoned with the coming of highly mechanized farming. The shallow soils are handicapped by low-moisture holding capacity, by excessive moisture in the spring and by insufficient depth for adequate root development.

Soils and Agricultural Capability

The diversity of Ontario soils is related to variations in climate and vegetation and to the complex pattern of surface features which resulted from the advance and melting of the ice sheets. The immature drainage pattern associated with glaciation accounts for the extensive areas of intrazonal-type soils (Fig. 1.3).

Along Hudson Bay where the drainage is impeded and weathering is slow, the soils are shallow, and acid in reaction. Throughout much of the Hudson Bay lowland, drainage is poor and bog-type soils are common. In these better drained areas, podzol-type soils have developed. A large part of the western part of Northern Ontario is characterized by shallow drift over Precambrian rock and by extensive bogs. Mature soils are restricted to the ridges of till and to the areas of deeper drift. The clay belts contain the bulk of the soils of Northern Ontario that are suitable for agricultural use. In these areas, more continuous stretches of deeper till and lacustrine sediments occur. However, wet, intrazonal soils covered with a variable depth of peat are common in the clay belts as well.

In the areas of Ontario where soil development is more advanced, three zones may be recognized. These are the podzol zone of the northern coniferous forest belt, the brown podzolic zone of the Rainy River District and the more southerly parts of the Shield with its mixed forest and the grey-brown podzols of the warmer Palaeozoic lowland of Southern Ontario with its deciduous trees. Podzolization is the dominant soil-forming process in the three zones. The process is most intense in the cool summer region of the north and least intense in the areas with the longest warm season and mainly deciduous-type vegetation. The brown podzols are intermediate and reflect the transitional nature of the climate and vegetation. In terms of the capability for agriculture, grey-brown earths may be rated highest and the podzols the lowest. The acid nature of the podzols,

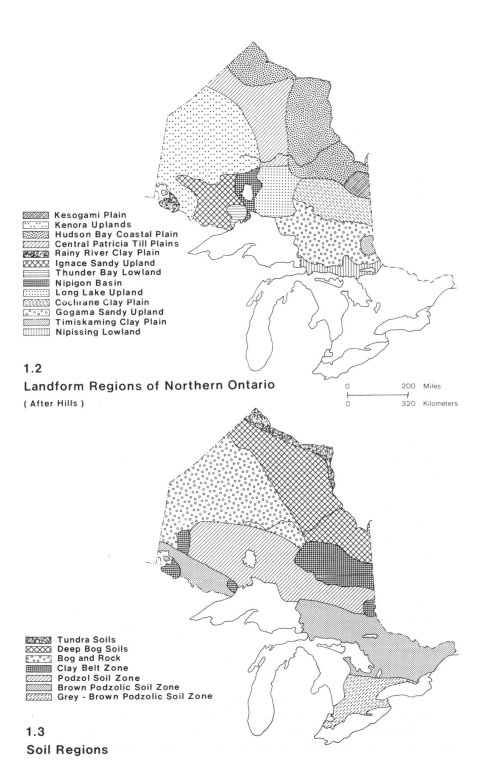

Kesogami Plain
Kenora Uplands
Hudson Bay Coastal Plain
Central Patricia Till Plains
Rainy River Clay Plain
Ignace Sandy Upland
Thunder Bay Lowland
Nipigon Basin
Long Lake Upland
Cochrane Clay Plain
Gogama Sandy Upland
Timiskaming Clay Plain
Nipissing Lowland

1.2

Landform Regions of Northern Ontario

(After Hills)

0   200   Miles
0   320   Kilometers

Tundra Soils
Deep Bog Soils
Bog and Rock
Clay Belt Zone
Podzol Soil Zone
Brown Podzolic Soil Zone
Grey - Brown Podzolic Soil Zone

1.3

Soil Regions

(After Ontario Soil Survey and Hills)

the slower breakdown of the vegetative matter, and the prevalence of peaty soils create problems in the utilization of the soils of Northern Ontario for commercial agriculture. Even the brown podzols are generally better adapted to forestry and recreational uses than to agriculture. About 20 per cent of the soils of the Palaeozoic lowland of Southern Ontario are light-textured and where the climate is favourable can be used for tobacco, tree fruits, and canning crops. About 40 per cent are heavy-textured and are best adapted to forage crop production and beef production or dairying. The remainder fall into the intermediate categories and have the widest range of adaptability.

The classification of soils, in terms of their capability for general and mixed farming has been completed for most of Southern Ontario and for parts of Northern Ontario. Maps have been published for most of Southern Ontario at a scale of 1:250,000 (Canada Land Inventory 1970). The soil capability classification system adopted by the Canada Land Inventory is an interpretative grouping of soil types which is based upon an evaluation of the physical conditions of the soil that influence agricultural land use. The mineral soils are grouped in seven classes according to their capacity or potentiality to produce general field crops. Each class is determined by the degree of total limitations to land use and the risks or hazards involved. For example, the top category or class 1 have soils which are highly suitable for cultivation and are not handicapped by high water tables, stoniness, shallowness, or any other physical characteristic which interferes with the use of tillage or harvesting equipment. Class 1 soils retain and supply sufficient moisture and nutrients for good yields of most crops, and they are not susceptible to water erosion.

The first three classes are regarded as suitable for most common field crops if properly managed. Classes 2 and 3 have certain limitations which do not apply to class 1. The fourth class is physically marginal for sustained arable farming while the fifth class is suitable only for permanent pasture and hay. Class 6 is capable of use only for grazing while class 7 is regarded as unsuitable for agriculture. While the soils in classes 1–4 are suited to cultivated crops, they are also suitable for permanent pasture. Soil areas in all classes may be suited to forestry, wildlife, and recreational uses. Generally speaking, the top classes include the better drained medium- and heavy-textured soils because these will produce high yields of most common field crops. However, there are a number of important cash crops such as tobacco, canning crops, and tree fruits which can be grown and indeed prefer the lighter-textured soils which may rate as low as class 4 in the National Soil Capability System. Organic soils which may be quite productive for certain vegetable crops when improved are not given a particular rating but have been designated by the letter O on the maps.

Miller and Hoffman have published three tables showing the acreages in each capability class for Northern Ontario, Southern Ontario, and the Near North (Miller and Hoffman 1970). These tables (Tables 1.2, 1.3, 1.4) include considerable acreages of land that have been urbanized or are now being used for non-agricultural purposes. At present, about 800,-000 acres of land are occupied by cities, towns, and villages. If population growth and urban expansion continue at the current rate, an additional 1,000,000 acres of agricultural land might be taken out of production by the year 2000. It would appear that Southern Ontario has approximately

Table 1.2   Soil capability classes for agriculture in Southern Ontario
(Includes land to the south of Lake Nipissing with the exception of the districts of Muskoka, Parry Sound, Nipissing and Manitoulin which are shown in Table 1.4)

Class	Million acres	% of total
1	4.9	20.3
2	4.7	19.8
3	3.1	12.9
4	1.7	7.1
5	1.4	5.8
6	1.8	7.5
7	5.2	21.6
	22.8	95.0
Organic soils	1.2	5.0
	24.0	100.0

Table 1.3   Acres of soil-capability classes for Northern Ontario

Class	Million acres	% of total
1	—	0.0
2	0.71	5.1
3	1.69	12.1
4	1.06	7.6
5	1.84	13.2
6	0.91	6.4
7	7.74*	55.6*
	13.95	100.0

*Acreages and percentages of organic soils are included in these figures.

Table 1.4   Acres of soil-capability classes in the 'Near North'
(Near North includes Muskoka, Parry Sound, Nipissing, and Manitoulin)

Class	Million acres	% of total
1	—	0.0
2	0.06	1.3
3	0.11	2.5
4	0.11	2.5
5	0.26	5.8
6	0.28	6.3
7	3.65	81.6
	4.47	100.0

12,000,000 acres of high quality arable land, about 53 per cent of its total agricultural area. The greatest concentration of high quality agricultural land is in Southern Ontario, and within this area there is some concentration in the most industrialized areas of the Golden Horseshoe and southwestern Ontario (Fig. 1.4).

Agricultural Land Classes
The author's work predates the Canada Land Inventory by 15 years (Reeds 1956). Figure 1.6 shows in a general way the distribution pattern of agricultural land capability classes for Southern Ontario. In viewing the map, the reader should keep in mind that land which has been assigned a low rating for general or mixed farming may be excellent for special crops such as tree fruits, tobacco, or market gardens. Although the classification is based primarily on texture of the soils, cognizance has been taken of other factors such as drainage, relief, climate, stoniness, and depth to bedrock. The first two categories include most of the best lands of Southern Ontario. Erosion is not a serious problem and relief does not impede the use of mechanized equipment. Third class land includes intermediate-textured soils, soil with steeper slopes and greater susceptibility to erosion and those with serious drainage problems. Fourth class land includes the lighter-textured and more open soils which have a lower moisture-holding capacity. This group also includes the bouldery soils, the shallow soils, some very poorly drained soils, and several steep phases. Where the climate is suitable, the sandy soils of this group may be quite productive for cash crops. The fifth class includes bottom land, marsh land, undrained bogs, rocky lands, and beach and dune sand. Much of this fifth class land may be ideally suited to forest growth, for wildlife preserves, or for recreational uses.

The map reveals the low proportion of high quality land in eastern Ontario. The northern parts of central Ontario also have a high proportion of marginal land. In the mid-sections of central Ontario, second-class land predominates with small scattered areas of other classes. To the south of the Interlobate moraine, a fairly high percentage of high-quality land prevails. This is also true of southwestern Ontario which benefits from its gently undulating topography, and the longer warm season. The reader should note that this is a very generalized map and each area shown in one class is actually a composite of all classes. The map indicates only the dominant class within the area delineated.

Climate and Agriculture
In an area as large as the province of Ontario, factors such as latitude, air-mass movement, location of storm tracks, location with respect to and distance from the Great Lakes, altitude, and relief features account for a variability in climate both regionally and locally. For example, the latitudinal extent of four degrees and the range of elevation of 1600 feet result in a variation of 26 days in the growing season in Southern Ontario. Other factors relating to microrelief, exposure, and soil characteristics are important to crop growth and forest reproduction.

The significance of climate to agriculture is presented on the map of climatic regions which combines temperature zones and moisture classes (Fig. 1.5) (Canada Land Inventory 1966). The first temperature zone coincides with the cash-crop farming area in Essex and southern Kent counties. Early vegetables are a specialty and two crops may be grown in one season. The second zone includes the Niagara Fruit Belt where winter temperatures seldom fall below 12° F below zero, and the tobacco belt of Norfolk County. A large area in Southern Ontario falls within the third zone where mixed farming, dairying, and livestock are more important than cash crops. Zone 4 has greater limitations with fewer crops including silage corn, hay, pasture, oats, barley and potatoes. With a 90-day frost-free period, the length of the growth season presents serious limitations to plant growth in zone 5. The frost hazard becomes even more severe in zones 6 and 7.

With respect to moisture, class G represents a fairly adequate moisture supply with rare droughts. However, yields are generally improved by irrigation in late summer. Class H has the most nearly ideal moisture regime in Canada. Surplus water, which often delays seeding and harvesting becomes an important problem in class K. In class L, midsummer droughts are very rare but wetness is a serious limitation particularly in the low-lying, heavy-textured soils.

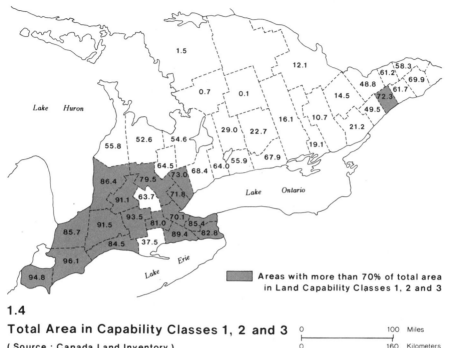

1.4

Total Area in Capability Classes 1, 2 and 3

(Source : Canada Land Inventory)

Areas with more than 70% of total area
in Land Capability Classes 1, 2 and 3

0   100   Miles

0   160   Kilometers

Region 1G and the Niagara Fruit Belt in 2G have the most versatile climate in Canada. The Essex-Kent area has the advantage of early springs, and the Niagara Fruit Belt with its mild winters, has the best and most reliable climate of any area in Canada for tender fruit production, including peaches and sweet cherries. For general farming, moisture class K rates close to H in the south but has to be rated lower in Northern Ontario because of the limitations imposed by a shorter frost-free period. Region 6L is rated below 6K in the Clay Belts of Northern Ontario. Zone 7L rates very low for agriculture but from a climatic standpoint is suitable for forest growth.

In the past, trial and error methods were important in the adaption of crops to climate. Climatic limitations are now well understood for most plant varieties. However, more research in microclimate needs to be undertaken before a more complete adjustment of crops and farming practices to minor climatic variations can be made. In many instances, such factors as the soil microclimate, hybrid plant varieties, as well as cultural and economic differences have an important influence on the significance of climate to agriculture. The interaction of physical and

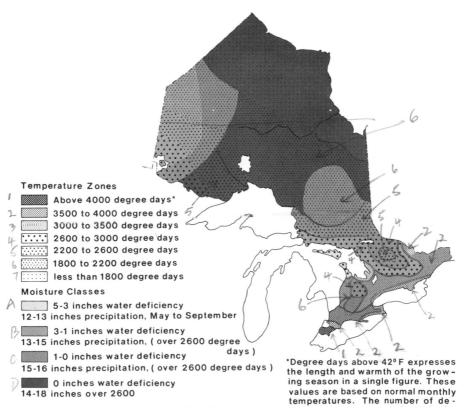

Temperature Zones

1 ▓ Above 4000 degree days*
2 ▓ 3500 to 4000 degree days
3 ▓ 3000 to 3500 degree days
4 ▓ 2600 to 3000 degree days
5 ▓ 2200 to 2600 degree days
6 ▓ 1800 to 2200 degree days
7 ▓ less than 1800 degree days

Moisture Classes

A ▓ 5-3 inches water deficiency
12-13 inches precipitation, May to September

B ▓ 3-1 inches water deficiency
13-15 inches precipitation, (over 2600 degree days)

C ▓ 1-0 inches water deficiency
15-16 inches precipitation, (over 2600 degree days)

D ▓ 0 inches water deficiency
14-18 inches over 2600

*Degree days above 42°F expresses the length and warmth of the grow- ing season in a single figure. These values are based on normal monthly temperatures. The number of de- grees above 42 were accumulated for all days between the dates of occurence of 42°F in the spring and in the fall. The totals for Ontario vary from 4250 degree days in Essex County to less than 1500 in the Hudson Bay lowlands (Canada Land Inventory,1966).

1.5

Climatic Regions for Agriculture

(After the Canada Land Inventory)

economic factors is frequently more important than the individual fac- tors themselves. For example, high dollar returns may encourage the pro- duction of a crop in areas of climatic risk. Favourable returns have encouraged tobacco production in the more northern areas of the province.

Recent Changes in Agriculture

Ontario's agriculture has been adjusting at a rapid rate to pressures of industrial and urban expansion. Farmers have been caught in a cost-price squeeze in which costs of production have risen more rapidly than prices received for produce. Increases in net farm income have not kept pace

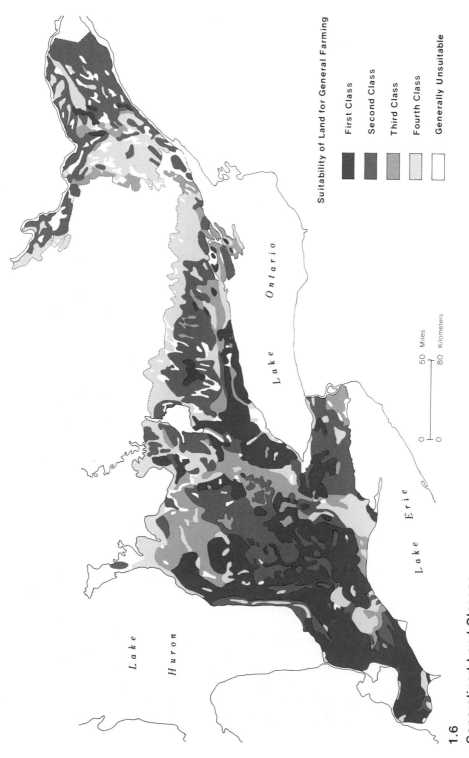

Suitability of Land for General Farming

First Class

Second Class

Third Class

Fourth Class

Generally Unsuitable

Lake Ontario

Lake Erie

Lake Huron

0   50 Miles

0   80 Kilometers

1.6

Generalized Land Classes

with increments in urban wages and other non-farm incomes. The adjustment has resulted in the abandonment of submarginal farms and more extensive use of land of mediocre quality.

The bulk of the reduced acreages has occurred in the districts of Parry Sound and Muskoka and in the counties of Haliburton, Hastings, Frontenac, and Renfrew where land quality is low. In the past, continuation of farming depended upon the availability of cheap labour and a very low total capitalization. The advent of high labour costs, a dwindling labour supply and an unwillingness to adjust to the new technology forced these operators out of business. Thousands of acres have been abandoned, while other farms have been enlarged and are being used mainly as ranches.

In the medium quality category, such as large areas in eastern Ontario and in the northern parts of southcentral and southwestern Ontario, much of the land is being used less intensively than prior to 1951. The small family farm with a limited acreage of high quality land can no longer compete in the highly mechanized and technological age. In these areas, much of the land that was formerly cultivated is now being used for pasture or is reverting to woodland. Under the Agricultural Rehabilitation and Development Act (ARDA), funds have been provided for drainage projects, for the amalgamation of farmsteads, and for the development of community pastures; even this assistance has not solved the problem of low incomes. In the more accessible sections, farms are being purchased by urban residents who use them for weekend retreats or as permanent homes.

While the bulk of the 'fallout' has occurred in the fringe areas of Southern Ontario and in Northern Ontario, where there is a high proportion of submarginal or marginal land, a large acreage of good quality land has been taken out of production in the highly industrialized regions. Areas which have experienced the greatest losses are near the industrial centres of Oshawa, Toronto, Hamilton, London, Kitchener-Waterloo, and in the Niagara Fruit Belt. The Lake Iroquois Plain between Hamilton and Toronto which prior to World War II was a highly productive agricultural region is now composed of an almost continuous string of low density housing, light industry, and shopping plazas.

In the mixed farming areas of the Niagara Peninsula, where over 80 per cent of the soils are classified as fair to good arable land and where climate and location with respect to markets are favourable, urban pressures have had a sharp effect on agriculture. Many of the small farms which could be operated profitably when costs of production were lower are no longer economical enterprises. As a result, many farms are being operated on a part-time basis, with the owner deriving the major portion of his income from an off-farm job. This is a characteristic trend throughout the mixed farming areas of the Niagara Peninsula, the northern parts of western Ontario, and in central and eastern Ontario.

Much of southwestern Ontario has experienced increasing specialization and greater intensification since 1951. The acreage in farms has remained relatively static, but yields per acre and gross incomes have increased dramatically. Production of cash crops has expanded greatly in the Lake Erie counties, while a more intensive agriculture has developed in the dairy farming and livestock producing areas of Oxford, Middlesex, Wellington, Waterloo, and Brant counties.

The changing patterns of agriculture in Ontario continue to reflect the interaction of physical and socio-economic factors. The author is convinced that man cannot completely overcome the physiographic controls and that the inherent capabilities of the land and climatic suitability will continue to affect production costs and land use. For these reasons, this chapter has emphasized the importance of understanding the natural landscape, its inherent characteristics and its use capabilities, and the need for ecological land-use planning. However, one must also bear in mind that technology is very important and that man can do much to make himself independent of the laws of nature. New varieties of many crops are being adapted to the less favourable climates as evidenced by the spread of grain corn northward in Ontario. With technological knowledge and efficient management, high yields of several types of crops are being obtained in soils which are far below the optimum physical quality. Other factors which offset the importance of physical land capability are land values, tax rates, transport costs, the availability of capital, the size of the operation and the market situation. In most cases, the viable enterprises are those which have taken advantage of economies of scale and are efficiently managed. For example, in the broiler chicken industry, the production of 140,000 birds per year is considered to be the minimum size for economical operation. Recent studies suggest that the most viable tobacco enterprises are those which have 40 acres or more in tobacco. The same trend is evident in practically every type of farming. Most of the small inefficient producers are expected to leave farming before the end of the century. Those technologically displaced farm people may not be able to find employment in the non-farm sector and constitute a social problem of considerable magnitude.

Table 1.5 summarizes the changes which have taken place with respect to area occupied, number of farms and farm population and projects the present trend to 1981.

The agricultural industry is changing so rapidly and is so strongly in-

Table 1.5   Changes in farming 1951–81

	1951	1956	1961	1966	1981 projection
Total area of farm land (acres)	20,880,054	19,879,646	18,578,507	17,826,045	16,767,000
Total farm population	702,778	683,148	524,490	498,075	350,000
Number of farms	149,920	140,602	121,333	109,887	75,000

fluenced by developments in the u.s., by the trend to free trade relations with the u.s., and by changes in the world market situation, that it becomes difficult to predict how much land will be needed in the future. However, until the picture becomes clearer, it is safe to assume that we should use our limited resources wisely and that we should attempt to conserve as much as possible of the best quality agricultural land for food production.

Site Regions and Forest Resources
Hills has subdivided the province of Ontario into 13 site regions (Fig. 1.7), with similar relationships between the development and growth of vegetation and landforms (Hills 1961). Regionalization criteria include the relief, the texture and petrography of geological materials, depth to bedrock, drainage conditions, and macroclimate. Site regions are not only landform and forest type regions, they are also biotic regions. As an integration of landscape features, the site region represents a biological productivity region having specific potentials. As such, it is a very useful tool for evaluating the potential of an area for forest productivity and for agricultural and recreational use, and should be considered as one of the criteria in land use and regional planning.

The pattern of site regions reveals a latitudinal arrangement of zones which reflect contrasts in temperature and moisture relationships from north to south (Fig. 1.7). The soil zones and forest divisions exhibit the same pattern. Vegetational types range from the tundra along Hudson Bay to the coniferous forest of central Northern Ontario, the mixed forests of the near north to the deciduous forests of the most southerly zone. Variations within these broad zones are related to differences in the local climate, landforms, and soils.

Forests are under the pressures of increasing demands for timber and pulpwood, the growing demand for minerals, the expanding interests of

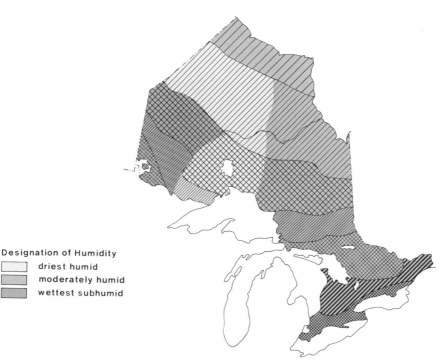

Designation of Humidity

	driest humid
	moderately humid
	wettest subhumid

Geographical Name	Climatic Designation
Hudson Bay	subarctic
James Bay	coldest of cold summer regions
Big Trout Lake	coldest of cold summer regions
Lake Abitibi	medium of cold summer regions
Lake Nipigon	medium of cold summer regions
English River	medium of cold summer regions
Lake Timagami	warmest of cold summer regions
Pigeon River	warmest of cold summer regions
Wabigoon Lake	warmest of cold summer regions
Georgian Bay	colder of the cool summer regions
Lake of the Woods	colder of the cool summer regions
Lake Simcoe-Rideau	warmer of the cool summer regions
Lake Erie-Ontario	warm summer region

A site region is an area of land within which the response of vegetation to the features of landform and the effective climate follows a consistent pattern. These regions have relevance not only for forests and their management but also for wildlife and to a certain extent for farming as well.

1.7

Forest Site Regions

(After Hills)

hunters and anglers and other recreational uses, and an increasing need for watershed protection. While forests may provide for all of these uses simultaneously, priorities will vary. Multiple-use plans based on sound ecological principles and safeguarding the best local use of the forests need to be worked out. The dominant function of forests in multiple-use regions is shown in Figure 1.8. Uses vary from the protection of bog landscape and wildlife in the extreme north to the scattered pulpwood producing areas of the Patricia and James Bay regions. Coming south, one enters the more productive timber and pulpwood areas. In the Shield of southern Ontario, forests provide an aesthetic landscape. They protect agricultural landscapes and water supplies in many parts of peninsular Ontario in addition to providing important green space in the highly urbanized areas.

Water Resources
Water is a vital natural resource. Certain uses do not involve the removal of water from its natural location. These include fishing, the conservation of wildlife and fish, swimming and other recreational activities, navigation, hydro developments, and the disposal of waste. Withdrawal uses of water include public water supplies, rural domestic and livestock watering uses, irrigation, and industrial uses.

Because of its population and industrialization, Ontario is by far the biggest water consumer in Canada; over 90 per cent is used for manufacturing and mining. In projecting the demand to 1990, it has been estimated on the basis of minimum flow, that Ontario will be using 31 per cent of its supply, a figure which is surpassed by one other region only, the Prairie Provinces (Cass-Beggs 1970). It is thus imperative that every possible effort be made by public agencies and private citizens to conserve Ontario's water supplies. It is not only a matter of conserving the quantity but of protecting the quality of the water. Already we have witnessed a drastic decline in the commercial fishing industry of Lake Erie because of pollution. Perhaps of even greater importance is the safeguarding of water because of the increasing interest in recreation. The inland waters of Ontario provide the major attraction for the tourist and recreational industry. Should this industry be seriously handicapped by pollution, the whole economy of Ontario would feel the effects.

Misuse of the Environment and Natural Resources
The tendency of our highly mechanized civilization has been to eliminate every trace of what was once primitive wilderness, to destroy the natural beauty of the rural landscape and to pollute the air, water, and soil.

The Dominant Function of Forests
in the Multiple Use Regions

Production of tolerant hardwoods,
hemlock, white and red pine, spruce
and fir.
Regions 8,9,14,18

Production of white pine, red pine
and spruce sawtimber; hardwood
and softwood pulp bolts and firewood.
Regions 5,10,12,13,16,17,24,28

Production of spruce, fir, poplar,
whitebirch, jack pine sawtimber,
poles and pulpwood.
Regions 15,19,20,22,23,26,29,30,31

● Major Pulp and Paper Mills

Present function, protection forest. Large areas of swamp with scattered areas having
potential production of spruce, fir and poplar pulpwood, largely difficult of access and
presently inaccessible except for hunting and fishing.
Regions 32,33,34,36

Protection of bog landscapes and wildlife values, scattered spruce, fir and poplar,
mainly of pulpwood dimension and largely difficult of access.
Regions 35,37

Protection of urban landscapes and water supplies.
Region 1

Protection of agricultural landscapes and water supplies. Scattered areas have highly
productive forest.
Regions 2,3,4

Protection of shallow soil landscapes with scattered areas of productive forest of all
types.
Regions 6,7,11,21,25,27

1.8

Multiple Use Regions

(After Hills)

Ontario's landscape reveals much evidence of man's reckless exploitation.

Entire species of plants and animals have disappeared; forests have been laid waste by uncontrolled cutting and fire, and top soil has been eroded away leaving the formerly productive land in a depleted, infertile state. Atlantic salmon which at one time abounded in Lake Ontario and its tributary streams have vanished. Sprawling super highways with their clusters of cloverleafs have gobbled up huge acreages of productive agricultural land. The great increase in demand for construction materials has resulted in the opening up of many new gravel pits and quarries which pollute the air with dust and noise and leave unsightly scars on the landscape.

Many of the rivers and Great Lakes shore waters have become seriously polluted with industrial wastes and domestic sewage. The air in most of the large cities is generally unhealthy. Large factory-type farms with high livestock populations are contaminating the air, water, and soil in their vicinity. Disposal of garbage is becoming an increasingly serious problem. Each person in Ontario is producing about one thousand pounds of garbage annually. A metropolitan centre, the size of Hamilton, produces enough waste in a year to cover 1000 acres to a depth of one foot. Overpopulation and uncontrolled economic growth could destroy the environment in the highly industrialized areas of Southern Ontario.

Demand for land for recreational uses is increasing very rapidly. Little space has been reserved on the Great Lakes shorelines for public use. Most of the inland lakes of Southern Ontario are becoming completely occupied by cottages. Ontario now has approximately 175,000 cottages, about 10 per cent of which are contributing to water pollution. In certain areas, building and sanitary regulations are lax. Cottage lots were approved in the past, primarily on the basis of their suitability for building, and septic tank operation. Little consideration was given to the size, shape, depth, and area of the lake in question; yet these are critical factors in determining the number of cottages a lake can accommodate without suffering extensive and perhaps irreversible environmental damage. Many of the cottage areas of Southern Ontario are overdeveloped. Drastic action will be required to clean up the inland lakes and to safeguard the cottage and tourist industry.

Population and pollution have been increasing at an alarming rate while the reserves of natural resources and the quality of environment have been declining. The emphasis on material gain and a high standard of living have accelerated this process. The time has come when we should be thinking not entirely in terms of increased productivity but of giving greater consideration to the way in which the increase will affect the environment. In the seventies, the improvement of the environment and the quality of life are goals which more people will be seeking, and economic growth which does not contribute to these goals will not be acceptable.

The provincial government has recognized the problem and has taken action to reduce and prevent pollution of air, water, and soil, to control cottage development, to regulate surface-mining operations, and to emphasize regional planning that stresses stricter environmental and ecological controls. In addition, an effort is being made to encourage a more even distribution of economic growth throughout the province, to assist each region to develop its potential resources and to separate out the highly urbanized areas with rural zones and park lands. In this programme, preservation of the best quality agricultural lands, protection of the aesthetic qualities of the rural landscape, and provision for recreational needs should be important objectives. If we are to improve the quality of life, every effort must be made to conserve the living resources and to protect the quality of the natural environment. If Ontario is to remain the prosperous place it has been, each individual must make an effort to clean up the environment and be willing to pay the costs involved.

Summary

In spite of the increasing importance of technology in primary resource industries, the present pattern of development continues to reflect the effects of the physical environment. Optimum areas for agriculture coincide fairly closely with the industrial heartland. The agricultural hinterland has great potential in its water, mineral, forestry and recreational resources. The critical problems facing the southern part of the province relate to the control of pollution, the acquisition of park lands in strategic areas for public use and the preservation of a portion of the best agricultural land. Conservation and wise use of natural resources and their development for the benefit of local residents should be one of the goals for Northern Ontario. Ecological land-use planning is the best method of accomplishing these objectives. In all parts of the province, the use of land for living space and for recreational purposes may become more important than the traditional function of economic production.

2 Settlement

R.L. GENTILCORE

The entry of European populations into an uncertain wilderness, the 'taming' of that wilderness and the creation of a new geography have been recurring themes in the settlement of North America. In Ontario, the sequence began late in the eighteenth century, establishing, within the next 100 years, the framework in which the province's commercial-industrial society operates today. Throughout the period, settlement sought accommodation with changing conditions. The main trends of this accommodation are outlined in this chapter.

INITIAL SETTLEMENT

In 1825, the province's population was 152,000. It extended along the St Lawrence River and the north shore of Lake Ontario and into the Niagara Peninsula. Westward, decreasing numbers were thinly strung out along the north shore of Lake Erie (Fig. 2.1). The main features of the distribution – the long distances covered, the attachment to water and proximity to the United States – were a direct result of the origins of the settlement and the way in which it was organized.

Settlement had its beginnings in the revolt of the American colonies. The first settlers were American 'loyalists,' rejected by the new nation because of their sympathy with the British Crown. In their wake came other Americans, looking for cheap land. The main points of entry for both migrations were key locations whose strategic value for movement had been recognized as much as 200 years earlier. The French had built forts at Kingston, Niagara, and Toronto, control points in their conduct of the fur trade. The first parts of Ontario to be settled were undeveloped but not completely unknown.

The first concern of the British was to establish a colony able to meet the threat of an independent, expanding United States. Military policy directed the settlement process: the organizing centres were military posts; soldier settlements were encouraged and only the Crown could secure and distribute land. Individual initiative and choice were submerged by the urgencies of the day.

The opening of land for settlement began with the alienation of Indian claims by the Crown. The next step was the land survey which laid out more or less rectangular townships and divided them into concessions and lots. Since surveys tended to precede occupance, their effect on the subsequent landscape patterns of rural Ontario was widespread and persistent. Born of the survey were such features as the alignment and spacing of farmsteads along the fronts of the concessions, the variation in land use between the fronts and the backs of the concessions, the rectangular shapes of fields and properties and allowances for roads at regular intervals (Gentilcore 1969).

The building of roads was another activity governed by military considerations. Under the newly created province's first governor, John Graves Simcoe, two main routes, Dundas and Yonge streets, were planned, westward and northward from the western end of Lake Ontario (Fig. 2.2). Their effect on settlement is indicated by the distribution of population in 1825. The effect of the Simcoe settlement plan on urban growth would become more obvious in subsequent decades, but its influence was evident even in 1825. Toronto itself was a creation of the plan as well as the smaller places beginning to appear along Yonge Street and to a lesser extent along Dundas Street or routes paralleling it (Kirk 1949). The plan's great achievement was that it directed settlement inland, away from the American frontier. It was the most active force countering the attachment to water.

Given the government's efforts to direct settlement, one may ask to what extent the results accorded with the physical geography of the occupied lands. A comparison of maps (Figs. 1.6 and 2.1) is instructive. Occupance was indiscriminate of physical conditions. It took place on all types of land surface, both along the lake shores and in the interior. In the initial stages of settlement, environmental advantages, even when perceived, were outweighed by other considerations such as accessibility and government direction. As settlements matured, the situation changed and choice and initiative were increasingly directed to the physical components of a complex which included soil and drainage conditions, availability of fresh water, proximity to a navigable waterway, the nearness of neighbours, and access to roads, towns, and mills.

The sequence of trees, fields, and towns so well developed in the settlement history of other parts of North America is a difficult one to identify in Ontario. In the earliest settled areas, all these appeared at the same time. The most obvious examples were Kingston, Niagara, and York

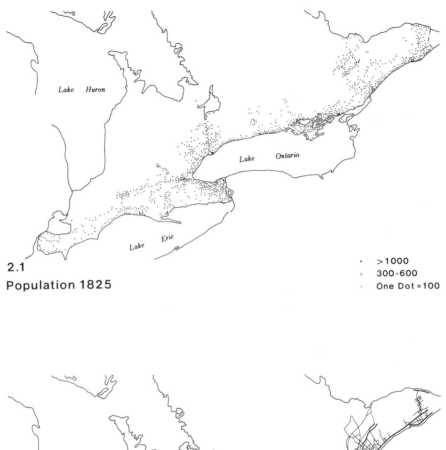

2.1

Population 1825

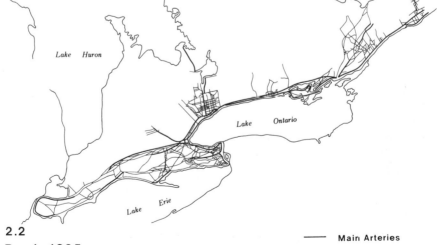

2.2

Roads 1825

Main Arteries

Others

whose association with heavier population densities and developed land routes is apparent on the two maps already cited (Figs. 2.1 and 2.2). The three places served as early organizing centres for the handling of settlers. They were political creations, within the framework of the Simcoe settlement plan, geared to the military needs of the new colony. All occupied sites on which the French had once built forts. They were on water where lake and river met and each was close to the United States.

The oldest of the towns, and the largest in 1825, was Kingston. Once the site of the French Fort Frontenac, it was chosen as the location for a major fortification facing the Americans. By 1810, with a population of 700–800, it was Upper Canada's most important naval base and garrison town (Preston 1959). It also became the province's undisputed commercial centre; agricultural products were collected here for distribution to other forts in Upper Canada. In 1825 its population was close to 3000. Contributing to its growth were a number of geographic advantages: its location at the junction of lake and water, its good harbour, and its well-settled hinterland. Particularly important was its position between Upper Canada and Montreal at a time when the province's business, trade, and financial operations were in the hands of Montreal merchants.

Like Kingston, Niagara was also a 'new town,' clustering around a fort built by the British to replace Fort Niagara which had been lost to the Americans. Its location on the most active frontier zone in the province led to its choice as the official provincial capital from 1791–5. The function continued, somewhat diminished, for another decade while facilities were being established in the new capital of York. Niagara's port was second only to Kingston's, deriving its importance from a strategic location where the Niagara River portage meets Lake Ontario. Its role as an entry point, particularly before 1812, led to a dense population in its hinterland. Although it declined in importance when it lost its role as capital, Niagara continued as the administrative centre of the Niagara peninsula, an important military post and a port complementary to Kingston on the Lake Ontario–St Lawrence route. Business firms in Montreal had branches here as well as in Kingston. Its population in 1825 was around 1200.

The youngest of Upper Canada's major urban places was York (Firth 1962, 1966; Kerr & Spelt 1965:13–54; Goheen 1970), built near the site of the former French Fort Rouillé. Simcoe chose it as a naval arsenal and the temporary provincial capital. He was impressed with its harbour and more particularly its location, removed both from Lower Canada (Quebec) and the American frontier. He saw it as a logical base for the movement of settlement into the interior. However, when the capital was moved in 1795, York possessed only twelve small cabins and a hinterland practically devoid of white population. In 1810, its population was less than 600, not a large figure for a provincial capital. Its slow growth reflected the lack of population in its hinterland, due in part to the large amount of land speculation by the small group of government officials in the new capital. The settlement, contrary to Simcoe's original plan, held on to its function as provincial capital. As this role grew, York grew with it. By 1825, its population was around 1800. It was rapidly catching up to Kingston, foreshadowing an early rise to a dominance it would never relinquish.

At the other end of the province, on the Detroit River, lay a fourth settlement, similar in many ways to both Kingston and Niagara. A vital link in the Great Lakes system, the area around the British fort at Detroit had a population of over 1000 in 1796, most of it on the right bank of the river. The British displacement to the other shore where a new fort was built, brought with it a modest population. Despite the functions of a military headquarters and a district centre in the same area, little urban development occurred. The location was too far removed from the main currents of population movement and trade. At the other end of Lake Erie, Fort Erie, though well located and active for a time as a fort and a crossing point, remained in the shadow of Niagara's administrative dominance. Another fort was established at Queenston. But the feature which contributed most to its growth was its location at the foot of the Niagara portage. Warehouses and fort installations were built before the fort, in 1789. At the peak of its development in 1810, Queenston had a population of 500–600.

In addition to the main places, geared to the political and military needs of the colony, there appeared a second set of places which had grown to supply local needs. Some were small ports, such as Belleville, Cobourg, Grimsby, and Port Talbot. Others functioned primarily as milling centres. Here farmers congregated, many of them coming from long distances and staying overnight or longer. Small inns, stores and perhaps a smithy might appear to serve their needs. Most of the milling places were shortlived. Some grew into villages, although this development was related more to the productivity and needs of the hinterland than to the presence of mills (Schott 1936).

A special feature associated with, and in some cases affecting, the distribution of population was its ethnic and religious composition. Here again, government policy provided the major impetus. Early townships were assigned in separate blocks to Catholic Highlanders, Scottish Presbyterians, German Calvinists, German Lutherans, and Anglicans, populations made up of 'loyalists' and disbanded soldiers. A whole county,

Glengarry in eastern Ontario, was set aside for a disbanded regiment of Scottish Highlanders, a settlement intended as a counterweight to American influence. The major inland settlement before 1825 was established by another distinct group, German speaking Mennonites from Pennsylvania. The desire of the community to stay together, to establish itself beyond the alien influences of other communities and to obtain large quantities of land cheaply led to the only instance of sizable settlement detached from water (Fig. 2.1).

Despite what might appear to be a mosaic of nationalities, spotted throughout the occupied areas of the province, the most important feature of Ontario's population composition was its American-British foundation. The 'loyalists' who were forced to seek refuge in Canada ensured a British presence in North America. With them, Ontario came into existence as a society with the will to remain distinct from the American experience. At first, it appeared that the 'loyalist' element would be submerged by other Americans who followed them in larger numbers. In 1810, eighty per cent of Upper Canada's 75,000 population was American: only one-quarter of these had been 'loyalist.' Evidence of American influence was everywhere. The log houses were American in origin. Many of the mills, blacksmith shops and taverns were owned by Americans. In the schools, set up after 1807, many of the teachers and the books they used were American. The expansion of religious groups such as the Methodists and other nonconformists was an American contribution. But the war with the United States, which broke out in 1812, brought this immigration to a halt. In addition, the need to defend the new community against American attack strengthened the allegiance of the new settlers to their adopted land. After 1814, there was no doubt that Ontario was something different from the United States, a divergence to be strengthened by subsequent immigration from the British Isles.

SETTLEMENT AT MID-CENTURY

Population at mid-century had grown to 952,000. The densest areas remained attached to water bodies, particularly along the St Lawrence River and Lake Ontario and in the Niagara Peninsula. The main changes were an infilling of numbers and their extension beyond the settlement frontier of 1825 (Fig. 2.3). The extension of trunk roads, such as those from Toronto northward and from Hamilton westward to serve new areas, accompanied the shift to the interior.

The building of roads and other transportation improvements were witness to the government's continuing concern and involvement with

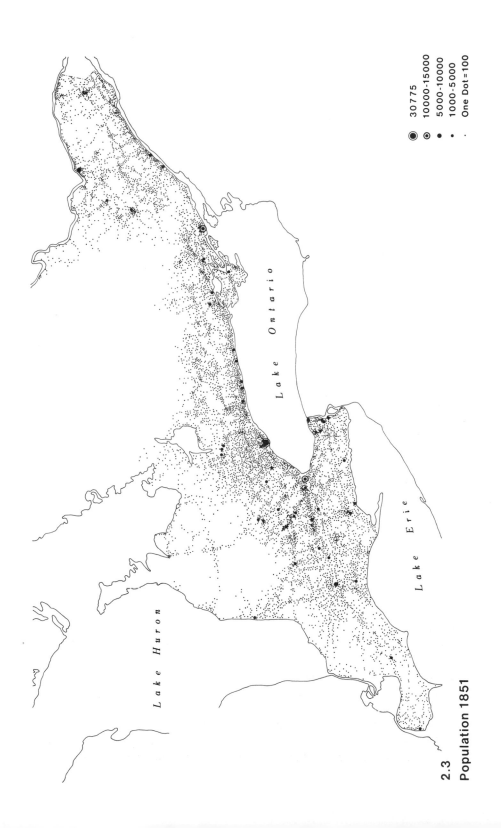

2.3

Population 1851

Lake Huron

Lake Ontario

Lake Erie

30775

10000-15000

5000-10000

1000-5000

One Dot=100

settlement. Following the War of 1812–14, the assistance given to earlier settlers from the United States was replaced by government schemes to encourage and assist groups from the British Isles to settle in Upper Canada. The tide of voluntary migration to North America was being drawn to the United States, by-passing British North America. Upper Canada, still sparsely settled, had to employ special means to attract some share of the migration.

In the early 1820s, groups of northern and southern Irish were brought to Lanark County in eastern Ontario, in an area which previously had received large numbers of Scots. A second scheme, under the direction of Peter Robinson, settled approximately 2000 southern Irish in another interior area, north of Rice Lake (Guillet 1957). Two other settlement schemes received their impetus from the Crown but were carried out under private auspices. The earlier one, under the direction of Thomas Talbot, began inconspicuously in 1803. Although intended to promote British settlement, Talbot's scheme could only survive at first with American participation. By the 1820s, however, sizable numbers, both from Great Britain and from other parts of the province were involved, taking up land along the Lake Erie shore from Long Point to the Detroit River (Coyne 1908). A second scheme involving government participation in the initial stages was carried out by the Canada Company to whom the Crown transferred 1,400,000 acres of Crown land, plus a large area known as the Huron Tract, embracing the present counties of Perth and Huron. Favourable government terms underscored the involvement of the Crown. The company reciprocated by vigorously promoting settlement. The high migrations of the 1830s and 1840s contributed to the success of the company and justified both its policies and those of the Crown.

In each of the examples cited above, the success of the settlement bore some relation to the degree of planning and care which went into the undertaking. In the case of the Lanark settlements, an area of refractory terrain was settled over a period of ten years, with little regard to physical suitability. By 1830, approximately one-third of the original Irish settlers had left for western Ontario and the United States. By mid-century few traces were left of any of them. The Kawartha scheme was successful. Not only did settlement persist but it also spawned the growth of a major regional centre. The town of Peterborough began as a depot for the Robinson settlement. The area, a drumlinized till plain with a variety of productive soils, was well chosen. The rolling landscape also proved an aesthetic attraction. A small number of highly literate settlers came together here and eventually produced a modest regional literature unique for the province and the times. (Among the authors who wrote about the

area were Catherine Parr Trail, Anne Langton, and Susanna Moodie.)
Accessibility to the earlier settled Lake Ontario shore enabled it to take
advantage of well-established ports of entry such as Cobourg and Port
Hope.

In both the Talbot and Canada Company operations, the taking up of
land was closely supervised to ensure its occupance and improvement.
The building of roads was a basic feature of both schemes. The Talbot
Road (Fig. 2.2) which became the main land route for southwestern
Ontario, was built and maintained by Talbot settlers as part of their settlement duties. The rigid enforcement of these duties distinguished the settlement from regions under governmental control. In the Huron Tract, the
building of roads and mills was the responsibility of the Canada Company.
The main axis was a planned route, the east–west Huron Road from the
eastern boundary of the Tract to Goderich, its designated capital. Following soon after was the north–south London Road, tieing settlement
developments to the growing populations around the London area. The
alignment of interior settlements along both roads is a feature of the
distribution of population in 1850 (Fig. 2.3).

The increasing role of private groups such as the Canada Company in
road building enabled the Crown to turn its attention to other transportation improvements. The most attractive was the canal. Canal building was
spurred on by American activities, particularly the opening of the Erie
Canal in 1825. Among a number of desirable connections, priority was
assigned to that between Lake Ontario and the Ottawa River, as the most
desirable for military purposes. The canalization of the Rideau was to
constitute the most expensive military undertaking by the British in North
America. Unfortunately, it yielded little return. The route was a round-
about one, of minor use for trade and for attracting settlement. Even its
military function was minimal. In contrast, the Welland Canal was conceived as a commercial venture, not a military one. Built with private
capital and opened in 1829, it was later purchased by the government.
Connecting Lakes Erie and Ontario, the Welland Canal gave access to the
upper lakes and drew settlement by immigrant labourers. It also facilitated
movement of population to western Ontario, by feeding routes westward
from Hamilton.

Increase in population and improvements in navigation, in turn, affected
production and export of Ontario's main commercial products, wheat and
timber. Wheat had steadily increased in acreage and production to the
1850s. It was easily the dominant crop as it has been from the early years
of settlement. However, the province's most important trade commodity
was timber. From 1830 to 1850, the value of timber exports regularly

exceeded that of all agricultural exports. The trade was promoted by government policy dating back to the substantial preferences given colonial timber in the early 1800s. The largest trade developed on the Ottawa, at the only place in Ontario with a fast flowing river leading directly to a major seaport and, as it turned out, in an area better suited to lumbering than to agriculture.

Urban Places

In 1825, only three urban places had populations of over 1000. In 1850, the number had increased to 38. Early growth was tied to immigration and expansion of farm populations. But increasingly, after 1835, urban development was related to the export economy based on wood, wheat, and flour. The cluster of urban places around the western end of Lake Ontario derives from their port functions and establishes a feature of urban distribution that survives and grows in the century that follows. The two largest cities, Toronto and Hamilton, were here as well as seven of the province's eighteen cities with populations above 2000 (Fig. 2.3).

Dominating the urban geography, both of the cluster and the province, was the city of Toronto. With a population of over 30,000, it was more than twice the size of Hamilton, the next largest city. Much of this difference could be attributed to its function as the provincial capital. At the same time, it shared in developments common to other Lake Ontario cities. Its port was a good one and the hinterland contributed increasing quantities of wheat for export. In the 1830s it was a major entry point and dispersal centre for immigrants occupying land north and northwest of Lake Ontario. By 1850, the region looked to Toronto as its wholesaling and distributing centre. It was the province's major importing port, although only slightly ahead of Hamilton; in exports, it still lagged behind both Kingston and Hamilton.

At the centre of the cluster, at the western tip of Lake Ontario, a series of urban developments brought the young city of Hamilton to the fore as the province's second city. Settlement in the area had begun in the 1790s, spawning a number of market centres. The largest, Dundas, was an early milling centre with access to water and good road connections to a hinterland that included the Waterloo settlements and those along Dundas Street to the Thames River (Fig. 2.3). Hamilton's growth dates from its selection as a district centre in 1816. A port function was added with the cutting of a permanent channel through the harbour's outer bar in 1832. A boom followed. Wharves, warehouses, and grain elevators were built. As roads were extended westward, the new city became the distribution centre of the interior. With the aid of the Welland Canal, coal was brought from

Lake Erie ports in the United States: steam power facilities were built and served as a basis for industrial development. The displacement of Dundas was complete. A water power site with limited access to water had given way to a new centre on a better harbour with easy access to new sources of power, a change typifying what was occurring elsewhere in the province.

At the eastern end of the peninsula, displacement was caused by the Welland Canal. The new route replaced the Niagara portage. Niagara, once the trade and marketing centre of the peninsula declined in importance. Its docks and shipyards fell into disuse; it lost its military and, later, its administrative function. The other Niagara ports suffered the same fate. Their place was taken by St Catharines, at the northern end of the canal, now the main port and with its new water supply, the new milling and manufacturing centre of the eastern peninsula. The building of locks in the lower Grand River, as part of the canal construction associated with the Welland, led to the creation of ports and milling centres at Dunnville and Brantford, enabling them to effectively tap their hinterlands, displacing numbers of local centres (Burghardt 1969).

Away from the primary cluster, the remaining populated places were closely tied to water. Of the 18 places in the province with a population of over 2000, 13 were ports; of 12 outside the cluster, 8 were ports (Fig. 2.3). The leading city in this group was Kingston. Although overtaken in size by Toronto and Hamilton it was still the province's leading exporting port, reflecting the continuing importance of the St Lawrence outlet and the role of Montreal in Ontario trade. The situation benefited other St Lawrence ports below Kingston, including Brockville, Prescott, and Cornwall. By 1850, however, Canadian trade via the United States vied with direct trade by sea. Kingston's role was declining. It experienced a spurt in growth when the provincial capital was moved there in 1841. The transfer back to Toronto two years later dashed whatever chances Kingston may have had to remain a major urban place in the province. Next in prominence to Kingston, also on a major river, but growing rapidly was Bytown (Ottawa). Its expansion derived from the lumber trade and the canalization of the Rideau for which it served as construction headquarters. However, the real impetus to its continued growth was to come later, with its selection as the capital of Canada in 1867.

With increase in population numbers and their spread inland, a small number of interior towns had appeared by 1850. Most important of these was London, Ontario's fifth largest urban place with a population of 7000. Despite Simcoe's plan to locate the provincial capital here, there was only a small village near the site in 1825. A sizable rural population had been drawn to the area, much of it through the efforts of Talbot and the Canada

Company. In 1826, London was named district capital and Dundas Street extended to it. Well served by land routes, with roads to Sarnia, Goderich, Hamilton, and Lake Erie, it became the focal point of southwestern Ontario. Export of wheat and flour from Port Stanley, serving the London hinterland, made it the third ranking port in the province. Nearby, in the Waterloo area, the development of service centres had also lagged. After 1835, new populations arrived from Germany, attracted by the presence of a German-speaking community. The European Germans, largely townspeople, settled in some of the small existing villages and formed new ones, creating an atmosphere in which town growth prospered. The largest town in 1850 was Preston which had begun before 1835 on the only important water power site in the area. More significant was the increasing prominence of Kitchener, without water power, but like Hamilton, accumulating steam power facilities which would serve as a base for industrial growth. A third inland town was the headquarters of the Robinson settlement at Peterborough, highlighting the role played in opening the interior by another colonization scheme. The rapid settlement of the area produced early agricultural surpluses. The Trent River, tapping the timber resources of the Shield made it an important lumbering centre.

The growth of towns to 1850 was increasingly tied to their ability to serve as market and service centres. But this ability, by itself, was not sufficient either to initiate or to quicken growth. Other functions were necessary; particularly important were the port and administrative functions. The former has received sufficient emphasis; the rise of the latter is described in chapter 6. The division of the province into districts was accompanied by the selection of district 'centres' which soon became the leading central places for the area around them. Another urban function, manufacturing, had not attained major status. The only exception, perhaps, was shipbuilding, scattered among a large number of lake and river ports, from Prescott to Niagara, reflecting the role of the ports and the timber trade. For the most part, manufacturing served local needs. Saw mills and flour mills were the basis of a complex which might include potash works, tanneries, distilleries and breweries, fulling mills, and, later, iron works and small woollen mills. Those places equipped with large-scale milling facilities possessed advantages for growth, but always limited by what the local area could absorb. The best prospects for industrial growth were tied not to water power sites but to shipping facilities.

Finally, it should be pointed out that the growth of towns, much of it incipient, was only part of an overall urban pattern dominated areally by small service centres. The pattern was an unstable one, reflecting the instability of populations and the immaturity of transportation facilities. As these conditions stabilized, so did the urban pattern (Spelt 1955).

THE LATE NINETEENTH CENTURY

The distribution of population in 1881 (Fig. 2.4) indicates a continuation of the filling-in process noted in 1851. More striking, however, is the movement northward into what were virtually empty lands in 1851, the Queen's Bush between Lake Huron and Georgian Bay and the Canadian Shield. Two Ontarios had now appeared. To the south was a rich and expanding community, a land of well-kept farms and growing cities. Northward was another land, where the forest was still dominant, farming poorly developed, settlement sparse and cities few. Economically and physically, the edge of the Canadian Shield was being established as the province's most obvious cultural divide.

'Old' Ontario

Of a total population of 1,920,000 over 90 per cent were located in those areas of the province effectively occupied three decades before. 'Old' Ontario had reached a certain level of maturity. Despite the pervasive influence of the heavy British immigrations from 1830 to 1860, it had formed its own institutions and shaped an identity which was neither American nor British. Less radical than the United States, less conservative than Britain, it had become a community old enough to have its own character. Its population composition reflected its increasing stability. By 1900, over 80 per cent of the province's population had been born here. It had become a supplier of emigrants, to the Canadian West, to the United States, and to its own empty lands to the north and northwest. The economy remained commercial and agricultural. But important changes had taken place. In the 1850s, the railway age had arrived, bringing with it new conditions for town and rural growth, a spur to industrialization, and a boom psychology that was to permeate all sectors of society.

Ontario's first effective railway was built in 1853. It paralleled the best developed land transport route, Yonge Street, connecting Toronto with Bradford on Lake Simcoe. In the following decade, the settled part of the province was covered by a network of lines. Basic to the system were the through or 'trunk' lines, designed to carry heavy traffic for long distances. The first of these was the Great Western Railway completed from Hamilton to Windsor and Toronto in 1855. As an undertaking, it exhibited many of the features that distinguished railway building in the province. First the line followed, in part, a well-developed route, Dundas Street, now a gravel highway. Secondly, it was promoted by a particular locality, the city of Hamilton, whose position on the through route between two main entry points from the United States, effectively stamped out competition

2.4 Population 1881

86415
10000-35000
5000-10000
1000-5000
One Dot = 100

Lake Ontario

Lake Erie

Lake Huron

from other towns in its area. The route itself emphasized another accompaniment of railway building, the increasing attraction of the United States market for Canadian goods.

The leading product carried by the railway and the one most closely associated with its beginnings and early development was wheat. The relationship between the two predates the appearance of the railway in Ontario. The approach of American lines to the Canadian border increased the demand for Ontario wheat both for milling in New York State and for shipment in bond to New York City. Its availability in the province (and beyond) encouraged American participation in railway building. The railway, in turn, played a key role as Ontario farm production moved in increasing quantities to the United States, under the stimuli of the Reciprocity Act of 1854 and the United States Civil War (Jones 1946). The wheat crop reached new peaks in acreage and production in the 1860s and 1870s; in 1870, 85 per cent of all Canadian wheat came from Ontario. The climax of production was reached in the early 1880s when total yields were twice those of 1861. But continuous production without fertilization had begun to exact its toll. Yields declined in the older areas. Even then, the de-emphasis of wheat was slow. It took the opening of the Canadian west to remove the crop as a major feature in the economy of Ontario. Eventually, even on good soil, under the best of conditions, wheat was no longer profitable.

Mechanization in agriculture was one component of an increased emphasis in manufacturing also fostered by the railway. The increased use of steam power in the latter part of the nineteenth century raised coal to the forefront as a basic ingredient for industrial growth. Railway transportation made it available to many places on the new transportation network. Inevitably, however, the assets of larger places, better equipped for both receiving and sending out goods, were felt. The trend to concentration became more pronounced. Up to 1881, manufacturing showed many of the features of the preceding period: the proportion of the labour force in manufacturing employment was limited (only 29 per cent in 1881), and firms were small and scattered. Changes in each of these characteristics were pronounced after 1881. Indicative of the trend is one figure which may be cited: in places under 5000 population, the number of firms decreased from 32,150 in 1891 to 6540 in 1901; at the same time the number of employees, approximately 160,000, changed very little.

The basis for the industrial developments of the late nineteenth century began to appear in the 1860s. A harbinger of things to come was the rise to prominence of the agricultural implements industry. The agricultural boom at mid-century led to a demand for machinery, supplied at first from the United States. Under tariff protection, from 1858 to 1866, its manufacture was firmly established in Ontario, aided by the setting up of branch plants by American firms. Ensuing competition for market advantage led to increasing concentration. The amalgamation of firms from Beamsville, Newcastle, Brantford, and Toronto to form the large Massey-Harris plant in Toronto in 1879 was the best example of the process.

The continuing role of the state in promoting Canadian manufacturing was confirmed with the imposition of high protective tariffs, as part of the government's National Policy in 1878. But government participation extended beyond tariffs. For example, cash bonuses were paid to stimulate the production of pig iron from domestic ores. By 1900, six blast furnaces had been built in Ontario. Only one survived, at Hamilton, with the government intervening to save the industry by paying a bounty on foreign ores. Subsequent expansion was based on ores from Minnesota and was accompanied by a sequence of events which culminated in the amalgamation of a number of firms which had been operating in Montreal, Quebec, Toronto, Gananoque, Belleville, Brantford, and Hamilton to form the Steel Company of Canada in Hamilton.

The pattern of growth set by agricultural machinery and steel was followed by a wide range of manufacturing activity. Expansion in production, concentration in location and the appearance of American branch plants became major features of growth. Accessibility by rail and the increasing market in other parts of Canada, particularly the newly settled west, helped to sustain and promote production.

Urban Places

The distribution of urban places in the last quarter of the nineteenth century continued a number of features from the mid-century pattern (Fig. 2.4). The major 'cluster' around the western end of Lake Ontario was still dominant. Also evident, although less pronounced, was the alignment with water bodies. The major difference was the large number of urban places in the interior areas, particularly in the western parts of the province. The overall distribution showed a balance with total population that was lacking in 1851. The latter part of the century was marked by a substantial growth in urban numbers. From 1851 to 1881, the number of places between 1000 and 5000 population increased from 33 to 132; places over 5000 increased from 5 to 14. For the province as a whole, the urban population increased from 133,000 to 571,000, from 14 per cent to 30 per cent of the population.

Urban change was closely linked to the railway. Smaller places unable to attract a rail line were doomed to decline and even disappearance. The concentration of economic activity in larger places at the expense of the smaller ones was an immediate and lasting effect of railway building. For example, the ability of the larger ports to reach further into the interior led to a concentration of the wheat trade at a few places. By 1860, over one-half of all the province's wheat exports were being handled by Toronto and Hamilton. Many of the small lake ports which had handled wheat and other exports up to 1850 lost their bases for existence.

The early urban network had been oriented to water. After 1851, the railway replaced water as the main means of transportation, promoted the wheat and timber trade and later contributed to the rise of manufacturing. The major trends in resultant urban growth are pointed up by a comparison of Hamilton and Kingston. Hamilton became an early railway centre. It collected wheat and other produce from the Niagara Peninsula and southwestern Ontario and vied with Toronto in attracting grain shipments from the Upper Lakes by rail from the Georgian Bay ports. Stimulated by agricultural prosperity and increased market demand, Hamilton used its port and rail facilities to bring in iron and coal for a metal industry producing engines, stoves, furnaces, and finally, railway equipment. Facilities established in the 1860s and 1870s and the protection afforded by the federal tariff attracted American capital. Spurred on by an aggressive city administration which provided tax exemptions and free land, an iron and steel industry was established, the first in the province. The city's subsequent development was closely tied to a rapid growth in manufacturing. In contrast, Kingston, although the province's fifth largest urban place, still had no railway to the interior in 1880. Its wheat came by water and even that supply was being reduced by the diversion of the grain trade to the American market. Production from its hinterland was limited. Although it continued to function as a trans-shipment point, Kingston did not develop as a major distribution centre. With the opening of the St Lawrence canals in 1905, its decline continued.

The most important railway centre was, inevitably, the largest city and the provincial capital. With railways, Toronto was able to expand its contacts beyond its immediate hinterland to the whole population area north of Lake Ontario. Its own particular line, the Northern Railway to Barrie and Collingwood, enabled it to reach to the Upper Lakes and was a forerunner of its subsequent ties to western Canada and the Ontario northland. Its growth as a grain market led to the foundation of the Toronto Exchange in 1855, for trade in wheat, flour, and other agricultural products. By the 1860s, it was the largest exporter of wheat in Ontario. Timber from Georgian Bay made it the centre of Ontario's lumber trade. Increasingly, the import trade concentrated at Toronto, contributing to the city's increased importance as the wholesaling and eventually the financial centre of the province (Kerr & Spelt 1965: 74–96). Financial strength fostered more trade, attracting manufacturing and leading to further growth. In 1881, Toronto had 10 per cent of the manufacturing employment in the province; in the next three decades, the figure increased to 27 per cent. Population rose from 31,000 in 1851 to 86,000 in 1881 to 377,000 in 1911. Toronto's total dominance of the province's urban geography was firmly established.

The North

The increasing maturity of settlement in Southern Ontario was accompanied by a movement northward into the untried lands beyond the settled areas. The border of the Canadian Shield had been reached before 1850. Knowledge of some of the area dated to the 1820s; its drawbacks were known. Nevertheless, settlement did eventually move northward, following in the wake of a hungry lumber industry and urged on by an expansion-minded provincial government.

The first lands opened for settlement were the non-Shield areas of the Queen's Bush, south of Georgian Bay, occupied as soon as they were made available. From 1851 to 1861, populations in Bruce and Grey counties increased from 2800 to 28,000. In 1881, the total was 65,000. But special inducements were needed for the Shield. Here the government reassumed the role it had played in the southern part of the province half a century earlier. It built colonization roads, advertised for immigrants, provided free land, and sold land to a settlement company. Finally, a railroad was extended into the area, reaching Gravenhurst in 1875.

At first, settlement grew slowly as farmers contributed both labour and their clearance timber to the lumber industry around them. But eventually, timber was withheld from the settlers, by decree. Its exploitation could only be done separately, by licensed lumbermen. Timber-oriented settlers left the area. By 1911, all Shield townships were heavily dotted with abandoned farms (and abandoned timber limits). The Shield townships of Peterborough county which had achieved a modest population of 3700 in 1891, dropped to 3100 in the next two decades.

The most important results of the push to the north were its accompanying transportation improvements and the continued growth of the lumber industry. For the next three decades, the white pine of the Shield moved in large quantities to the older parts of Ontario and beyond them, to markets in the United States. Lumbering encouraged the growth of established towns such as Peterborough and Lindsay, along a major axis of movement, the Trent River system; it brought prominence to the

Georgian Bay towns, particularly Collingwood and Midland where the use of steam led to a concentration of saw milling activities and the resultant shipment of their products by rail to Toronto and Hamilton. Eastward, on the Ottawa River, the Ottawa–Hull area achieved new prominence, although its timber supply now came largely from Quebec.

The march of the timber industry led to the growth of a small number of towns as supply bases and mill sites. Orillia achieved prominence in the 1850s as the gateway to the northern areas. As timber operations spread over the Muskoka region, Bracebridge became its commercial centre. By 1871, a small amount of agriculture was being carried out in its hinterland, with hay and oats being produced for the timber camps. Town development duplicated parts of the southern pattern. The use of water power permitted the establishment of small industries including a tannery and a woollen mill. Transportation facilities included a colonization road and a railway line which arrived in 1885. Continuing prominence was assured with its choice as district centre and later, county capital. Another place which grew along similar lines was Parry Sound, the centre for timber operations east of Georgian Bay. In the Haliburton area, the small village of Haliburton remained the only survival of an ambitious colonization scheme.

The movement into the Shield had other repercussions. It led to a realization that not all parts of the province were the same, and that the settlement experience of the south could not be repeated here. The land was different; the soil was thin and timber resources limited. Removal by lumbering and destruction by fire laid waste immense areas. The extent of the damage led to a new assessment of the land. For the first time, the need to protect it, rather than promote its unhindered use, was recognized. In 1893, the government of Ontario set aside 2700 square miles south of the Ottawa River as a provincial park (Algonquin Park) where preservation and protection were to take precedence over production.

The northward movement had another important effect. It facilitated the entry and establishment of settlement beyond Lake Nipissing and the northern shores of Georgian Bay, in what is now officially 'Northern Ontario.' Entry had begun before 1881. By 1891, the population was 55,000; by 1911, over 200,000 people were scattered from the Quebec border to that of Manitoba (Figs. 2.5, 2.6). The settlements did not penetrate the Shield but hung precariously to its edge, in a pattern that was destined to persist. In the south, land for farming had been the backbone of occupance. The land was taken up and production from it moved to a network of service centres covering all of the old settled areas. It was different in the north. The hinterland was not a productive agricultural area but a

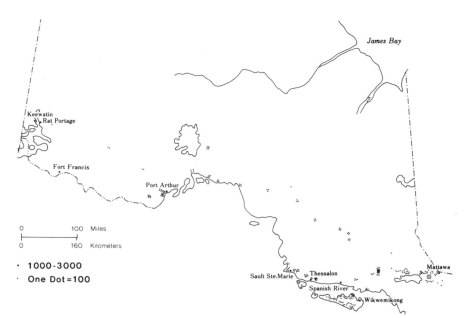

· 1000-3000
· One Dot = 100

2.5

Northern Ontario, Population 1891

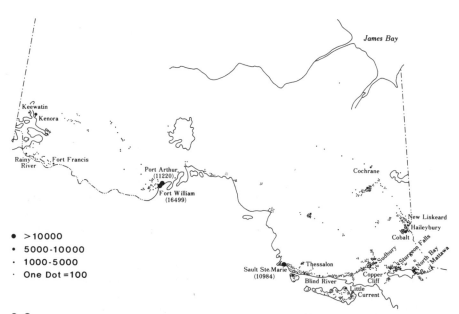

● > 10000
● 5000-10000
· 1000-5000
· One Dot = 100

2.6

Northern Ontario, Population 1911

supplier of timber and minerals. Settlement grew next to these resources; it was spotty and scattered, reflecting their distribution. Transportation was critical, as it had been in the south.

A new resource played a key role. In the south, some mineral wealth had been found, including surface iron ore, gypsum, petroleum, natural gas, and salt. But deposits were small and activity associated with them shortlived; they had little influence on settlement. In the north, minerals proved to be the lifeblood of many communities. Silver sparked the first interest in the Thunder Bay area. A substantial iron deposit discovered at Michipicoten in 1897 helped establish an iron and steel industry in the north. Nickel and copper production began in the Sudbury area in 1887. Gold drew populations westward to Kenora and at the other end of Ontario, northward to Porcupine.

Minerals did not operate in isolation. They were part of a complex that included water power, timber, transportation facilities, some agricultural achievements, and government participation. By 1900, the most impressive industrial establishment in the north was in Sault Ste Marie, embracing steel, hydro power, pulp and paper, and sawn timber. It was located on the edge of the Shield, at the point of physical contact with the United States and on water communication with the south. It was not yet on a transcontinental railway line, although this deficiency was remedied with the construction of the Algoma Central and Hudson Bay Railway, built to carry raw material to the pulp mill at the Sault. The town's population rose from 2000 in 1894 to 8500 in 1902, with an additional 3500 in the surrounding area (Konarek 1970).

Over 300 kilometres to the west, the Thunder Bay area, dependent first on mining and lumbering, became established as an important transfer point with the coming of the transcontinental railway and the building of grain elevators to handle wheat from the west. Another long gap separated Thunder Bay from Kenora and Fort Frances, both nurtured by lumbering and water power and both primarily oriented to markets nearby in the United States. Kenora, following the Thunder Bay pattern, surged ahead with the coming of the railway and the building of flour mills.

Along the Quebec border, population had moved northward via the Ottawa River and Lake Timiskaming. In contrast to other parts of the Shield, this area contained some good agricultural land, the 'Little Clay Belt.' A combination of lumbering and agriculture permitted a new growth of settlement. Populations in the districts of Nipissing and Timiskaming more than doubled in the decade after 1891, reaching 29,000 in 1901. The increases and the evidence of good soil farther north in the 'Great Clay Belt' spurred the Ontario government to action. In the tradition of its predecessors, it undertook the promotion of agricultural settlement on a grand scale. A colonization railway, the Timiskaming and Northern Ontario, was begun from North Bay in 1902. When it reached Cobalt the following year, the discovery of silver turned attention away from agriculture. The railway could not keep up with the mining frontier. Gold was discovered at Porcupine, leading to a concentration of population in the district by 1911. At the same time, Cochrane, at the junction of the incompleted line with the transcontinental Grand Trunk Railway, was emerging as a centre for the area farther north.

These modest settlement beginnings established a new element in the geography of Ontario. The minerals and forests of the north became an important component in the continuing economic development of the south. In particular, Toronto's growth as a financial centre, rivalling Montreal, derived from its control of northern resources. The complementary nature of the two regions put increased emphasis on transportation facilities, both in the establishment of new ones and in the heavier use of some that existed at the turn of the century.

Summary
The reconstruction of European settlement patterns in nineteenth-century Ontario elucidates the establishment of certain basic features in the geography of the province. Early settlement was strung out over long distances, close to water and to the United States. It was organized from centres selected by military authorities, so that from the beginning rural and urban settlement appeared together. The urban component grew as part of an export economy based on wheat, flour, and wood. Before 1850, a cluster of urban places, most of them ports, had been established around the western end of Lake Ontario. Even with the growth of manufacturing, the best prospects for growth remained tied to shipping facilities. The major features of the settlement pattern in the late nineteenth century were the persistence of the cluster, the alignment of population with water and land transportation routes, and the extension of settlement into the interior. In the north, a new community emerged, supplementary to the south and dominated by it. The two Ontarios and the concentration of urban population were creations of the nineteenth century destined to persist as major features in the contemporary geography of the province described in the following chapters.

3 The Economy

D.M. RAY

SPATIAL FORM: A DIAGRAM OF FORCES

The spatial form of economic systems reflects the action of forces such as the spatial flows of investment capital and migrants, the diffusion of innovation and the organization of decision-making during the growth of the system. System growth may be simple or structural. The former entails only an increase in numbers, but is limited by the level of organization of the system. Growth creates form, but form limits growth (Boulding 1953). Structural growth, on the other hand, entails the development and evolution of the form of a system; and it may promote continuing growth, with past growth locked in by a ratchet-like safeguard and future growth ensured by the institutionalization of innovation.

The relative importance of components may change during growth. In many cases, relative growth follows the law of allometry in which the ratio of the components, or of one component compared to the entire system, remains constant over the growth period (Naroll & von Bertalanffy 1956). The structural growth of economic systems, as measured, for instance, by the urban to rural population growth, and the growing diversification of occupations of the labour force to the size of the largest city, is itself allometric.

The growth and form of the Ontario economic system is composite in nature with heartland–hinterland, urban hierarchy and development axes comprising the dominant components (Ray 1969). These components may be related to centripetal and centrifugal forces acting: on a national scale to tie the Prairie and Atlantic hinterlands to the Ontario–Quebec heartland, on a regional scale to tie urban hinterlands to their urban centres; and finally between pairs of major urban centres to create development axes. A vestigial east–west structure of many economic characteristics remains in Ontario from centripetal and centrifugal forces operating at an intercontinental scale when Canada's economic development was related to the production of staple exports for Europe. The spatial structure of Ontario's economy is described below in terms of the three dominant components, their growth and the underlying forces.

THE HEARTLAND–HINTERLAND STRUCTURE

Heartland–Hinterland Contrasts

Heartlands are 'territorially organized subsystems of society which have a high capacity for generating and absorbing innovative change; peripheral (or hinterland) regions are subsystems whose development path is determined chiefly by heartland institutions with respect to which they stand in a relation of substantial dependency' (Friedmann et al. 1970). Heartlands are not as extensive as 'ecumenes,' which are generally defined to include the entire populated area, though the term ecumene too has sometimes been restricted to the most viable area of a nation (Hamelin 1968). Canada's heartland may be identified by its high market potential extending from Windsor eastward to Quebec City (Fig. 3.1). Thus defined it corresponds with the Southern Ontario and Quebec ecumene, but excludes the Prairie ecumene.

Heartland–hinterland contrasts are a pervasive element of the geography of Canada. Family-income levels decline progressively and significantly with distance from the heartland; so do market potential, the number and per cent of male labour force employed as craftsmen, and postwar immigration as a per cent of all foreign-born. Economic disparity, measured as the weighted difference between population potential and market potential, increases significantly with distance from the heartland (Fig. 3.2).

In general, two groups of characteristics have significant heartland–hinterland contrasts. First, there are contrasts in education, occupation, income, and housing characteristics that are primarily urban–rural in nature, for urbanization itself is concentrated in the heartland. Second, there is a distinctively heartland–hinterland group of characteristics which contrasts the relative emphasis on manufacturing in the heartland with a lumbering–fishing–mining economy and associated higher unemployment rates and greater economic disparity at the periphery (Fig. 3.3).

The Growth of the Heartland

Economic contrasts have tended, in general, to increase both within the heartland itself and between the heartland and hinterland in the century since Canada's confederation. The combined effect has been a growing disparity in industrial and urban growth between the 'Golden Horseshoe' extending from Toronto to Niagara Falls, and the rest of Ontario. Indeed,

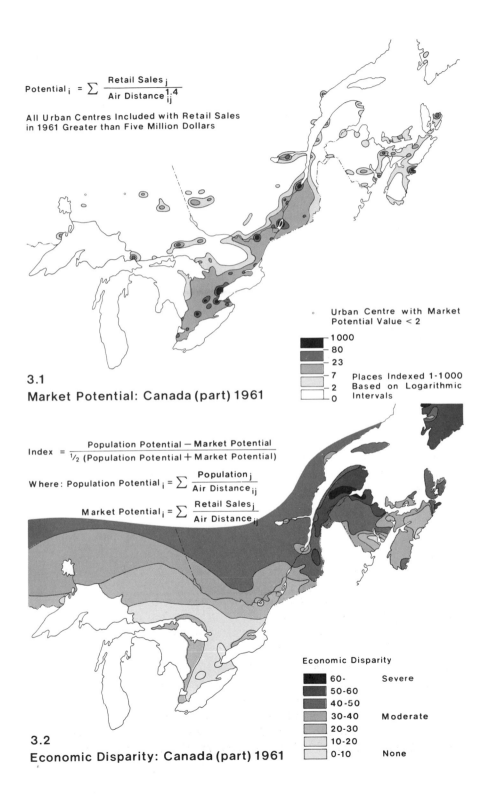

$$\text{Potential}_i = \sum \frac{\text{Retail Sales}_j}{\text{Air Distance}_{ij}^{1.4}}$$

All Urban Centres Included with Retail Sales
in 1961 Greater than Five Million Dollars

Urban Centre with Market
Potential Value < 2

■	1000
	80
	23
	7
	2
	0

Places Indexed 1-1000
Based on Logarithmic
Intervals

3.1
Market Potential: Canada (part) 1961

$$\text{Index} = \frac{\text{Population Potential} - \text{Market Potential}}{\frac{1}{2}(\text{Population Potential} + \text{Market Potential})}$$

$$\text{Where: Population Potential}_i = \sum \frac{\text{Population}_j}{\text{Air Distance}_{ij}}$$

$$\text{Market Potential}_i = \sum \frac{\text{Retail Sales}_j}{\text{Air Distance}_{ij}}$$

Economic Disparity

■	60-	Severe
	50-60	
	40-50	
	30-40	Moderate
	20-30	
	10-20	
	0-10	None

3.2
Economic Disparity: Canada (part) 1961

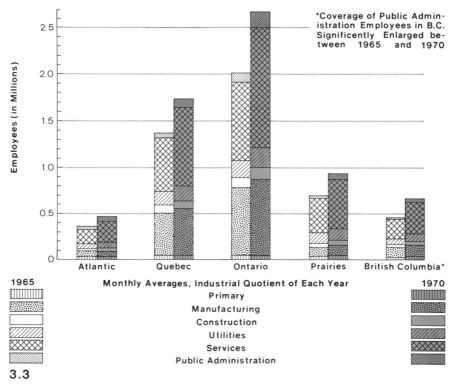

3.3

Industrial Development by Major Sector 1965, 1970

(Source: The Bank of Nova Scotia)

Toronto has grown to rival Montreal and has had the highest market potential in Canada since 1931, the first census year for which retail sales data are available; a third of all Canadian retail sales are made within a hundred miles of Toronto. However, Montreal's market potential is only a little lower than Toronto's and its population potential to date has always been higher.

The increasing provincial dominance of the Toronto metropolitan region, which is described in chapters 2 and 5, is evident in the changing provincial distribution of population and manufacturing employment in the century since confederation (Figs. 3.4, 3.5). Also evident is the growing disparity between the areas east and west of Toronto. Whereas urban centres were equally distributed east and west of Toronto in 1861, a century later there are four Southern Ontario metropolitan areas west of Toronto compared with only one to the east. Furthermore, in each of the

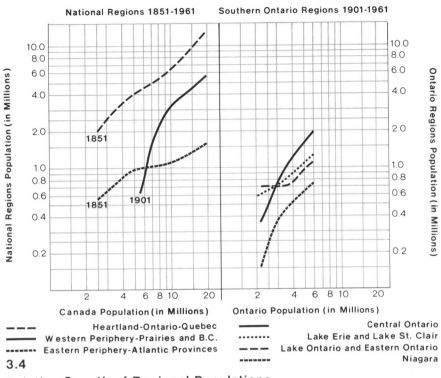

National Regions 1851-1961   Southern Ontario Regions 1901-1961

— — — Heartland-Ontario-Quebec	———— Central Ontario
———— Western Periphery-Prairies and B.C.	········ Lake Erie and Lake St. Clair
········ Eastern Periphery-Atlantic Provinces	— — — Lake Ontario and Eastern Ontario
	——— Niagara

3.4

Relative Growth of Regional Populations

urban size groups, 2500–5000, 5000–10,000, and 10,000–30,000, there are at least twice as many centres west of Toronto as east of it (Spelt 1968). The emergence of Toronto as Canada's largest centre in terms of market potential has thus been linked with the emergence of a distinct cluster of southwestern Ontario urban centres (King 1966) and a continuing expansion of the Canadian heartland toward the contiguous parts of the United States heartland in upstate New York and in Michigan. Indeed, the Canadian heartland has been portrayed as part of the 'Great Lakes Megalopolis' extending from Chicago to Quebec City, and linking the midwest to the 'eastern megalopolis' along the Atlantic coast (Doxiadis Associates 1969). This megalopolitan concept implicitly recognizes the increasing impact of interaction between the Canadian and American heartlands on the spatial structure of social and economic characteristics in Southern Ontario.

A disparity also occurs in agricultural development between the southwestern and eastern regions of the heartland that is related to eastern

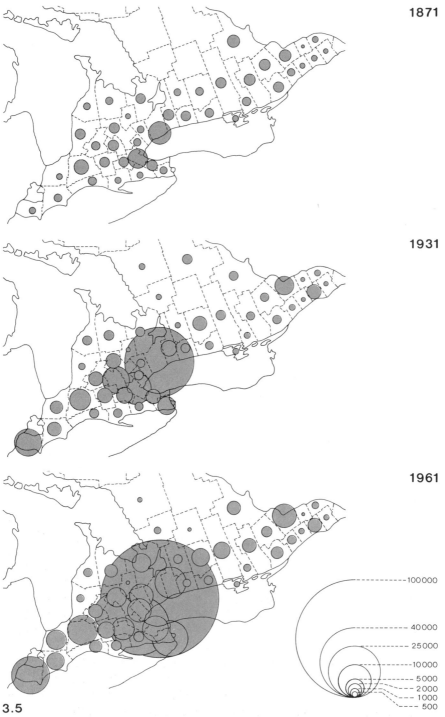

1871

1931

1961

---- 100000

---- 40000

---- 25000

---- 10000

---- 5000
---- 2000
---- 1000
---- 500

3.5
County Manufacturing Employment, 1871, 1931, 1961
(Source: Economic Atlas of Ontario)

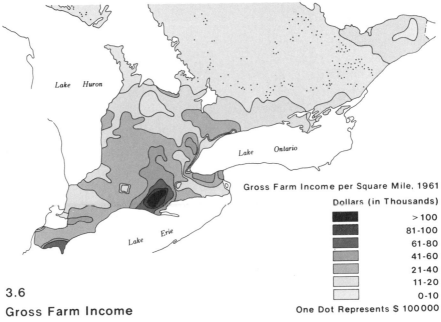

3.6

Gross Farm Income

(Source: Economic Atlas of Ontario)

Gross Farm Income per Square Mile, 1961

Dollars (in Thousands)

	> 100
	81-100
	61-80
	41-60
	21-40
	11-20
	0-10

One Dot Represents $ 100 000

Ontario's less favourable physical environment and smaller urban markets for agricultural products. As described in chapter 1, agricultural productivity increases from the north and east toward the southwest; a longer growing season here permits the production of higher-valued cash crops so that some net increases in farm acreage are occurring despite losses due to urban expansion and highway building (Fig. 3.6). Conversely farm acreage losses have been greatest to the north and east where the prevalent shallow and rocky soils, unsuited to mechanization, are being abandoned or reforested. Studies of agriculture in eastern Ontario suggest that about a quarter of the land presently farmed is better suited to forestry and that a third of the farms are not, in fact, viable (Noble 1965). The average net income from all sources for a representative sample of full-time farmers in eastern Ontario in 1961 was found to be $2528, less than half the provincial average income for wage earners, and too low to support an adequate standard of living. The concentration of Ontario farm poverty in eastern Ontario thus combines with that region's lagging industrial and urban development to partition the Canadian heartland into two: the urban heartland of southwestern Ontario which is contiguous to the United States heartland, and the Montreal-centred heartland of Quebec.

The Heartland–Hinterland Process

Canadian growth and development have not been elaborated in the literature specifically in terms of a national heartland–hinterland process.
Nevertheless, the staple export theory, a Canadian contribution to regional
economic development literature, applies the concept at the international
scale and argues that, 'the economic history of Canada has been dominated
by the disparity between the centre and the margin of western civilization'
(Innis 1967). Furthermore, Robinson develops the national heartland–
hinterland theme in his examination of the staples of the Canadian Shield
and demonstrates an outward-moving distribution pattern of resource
developments from central cores in the southern Shield. The first exploitation of natural resources, as described in chapter 2, took place along
the southern edge because it adjoined the areas of relatively high population density and economic activity in the St Lawrence Lowlands, and was
near Canada's two largest urban and industrial centres, Montreal and
Toronto (Robinson 1969).

The heartland–hinterland process relates economic development to
regional endowment and market accessibility. Continued and self-sustain-
ing growth depends on circular and cumulative causation in which a region
successively attains the threshold for the internal production of a wide
range of goods and services and thus achieves the associated economies of
scale. Centripetal forces add leadership in finance, education, research,
and planning to the initial advantages of the heartland. Secondary manufacturing and service activity then gravitate towards it, leaving hinterland
areas reliant on primary industries which tend to play a diminishing role
in national economies. Centrifugal forces, which reduce heartland–hinterland contrasts, include: the benefits that growing markets and improving
technology at the centre can have on localities in the hinterland; the protection afforded hinterland industry by distance from the heartland; and
the increasing congestion of the heartland combined with special amenities
which parts of the hinterland may have to offer. The operation of the
centripetal–centrifugal forces in Ontario are described in chapters 2 and 5.

Heartland–hinterland contrasts reflect centripetal and centrifugal forces
in which, at the national scale, the centripetal forces are dominant. The
Canadian heartland–hinterland gradient in population potential, which
was decreasing from 1871 to 1901, has increased steadily ever since,
irrespective of depression or war (Table 3.1). Hence, Toronto and southwestern Ontario can be expected to have an increasing proportion of
Canada's population unless the balance of the centre-periphery forces
changes.

Table 3.1   Population potential gradients from Halifax and Montreal, 1871–1961
(Koenig 1971)

Year	Gradient of population potential from		Coefficient of multiple determination (R^2)
	Halifax (b_1)	Montreal (b_2)	
1961	0.114	−0.383	0.59
1951	0.093	−0.363	0.60
1941	0.095	−0.360	0.63
1931	0.108	−0.353	0.62
1921	0.082	−0.327	0.63
1911	0.098	−0.278	0.67
1901	0.065	−0.283	0.57
1891	−0.004	−0.319	0.53
1881	−0.091	−0.367	0.54
1871	−0.129	−0.388	0.55

NOTES

1   The population potential gradients are given by the multiple regression coefficients in the equation:

$$\log Y = a + b_1 \log X_1 + b_2 \log X_2,$$

where Y = population potential for each census division for given census year,
X_1 = distance of each census division from Halifax,
X_2 = distance of each census division from Montreal.

The equation is computed for each census year separately.

2   Distance from Halifax is a surrogate measure of heartland–hinterland forces at an international scale. Distance from Montreal, the point of highest potential 1871 to 1961, is a measure of these forces at the national scale. By Confederation (1867), the east–west gradient was already weaker than the national heartland–hinterland gradient and was reversed from negative to positive at the turn of the century. Since then, population potential has tended to increase westwards from Halifax, and decrease outwards from Montreal at rates which have increased fairly steadily. The settlement in the Prairies, which was particularly rapid from 1896 to 1913, is an important factor in explaining the decrease in the slope of population potential from Montreal between 1871 and 1911. The growth of Toronto and the decreasing primacy of Montreal may be a contributing factor.

It is more difficult to assess the heartland–hinterland gradient in personal income. Analysis of personal income per capita for the five major Canadian regions over the period 1926–65 shows stubborn regional disparities with Ontario's figures, double those of the Atlantic Provinces region (Economic Council of Canada 1965). If the disparities are measured using retail sales at the county level, however, important regional disparities appear within Ontario with the Toronto–Hamilton and London areas having the lowest disparity index and disparities increasing with distance from southwestern Ontario (Fig. 2); furthermore, regional disparities may have narrowed across the nation from 1931 to 1961.

Much more research is needed before the Canadian heartland and hinterland can be fully defined, and the strength of the centre–periphery forces adequately assessed. The evidence available indicates that centripetal forces are powerful agents operating increasingly in favour of the continued concentration of urban growth in Toronto and southwestern Ontario. At the same time there is no evidence to suggest that an increasingly concentrated pattern of urban growth is leading to increasing regional disparities.

THE URBAN HIERARCHY

Diffusion and Urban Fields

Heartlands and hinterlands are articulated into a national system through the hierarchy of metropolitan and urban centres: 'In particular, large economic establishments – such as corporations and financial institutions – have tended to congregate in the metropolis, where policies are shaped and from which decisions are diffused through successively smaller cities to all corners of the country. In reverse, funds, materials, and people move from the hinterland to regional cities and on to the metropolis' (Kerr 1968, p. 531). Innovation and decision-making appear to diffuse (a) centrifugally from heartland metropoli to those in the hinterland, (b) hierarchically from larger urban centres to smaller centres, and (c) outwards from urban centres across their urban fields. At higher levels of economic development, the rate of diffusion can be expected to be more rapid, and distance–decay less marked, thus increasing the prominence of hierarchical diffusion.

The process of spatial diffusion produces interdependencies and disparities between metropolitan centres and their hinterlands that repeat at a regional scale the heartland–hinterland interdependencies and disparities found at the national scale. Furthermore, the regional heartland–hinterland systems can be expected to gain increasing prominence with the continued development of the Canadian economy.

The Manufacturing Hierarchy

The dominant function in most larger Canadian cities, particularly heartland cities, is manufacturing. Maxwell, utilizing the Ullman–Dacey technique, found that all cities in Southern Ontario, except Ottawa, had most of their basic employment in manufacturing (Maxwell 1965). He therefore subclassified manufacturing cities into manufacturing I, in which

Table 3.2   Functional classification of Canadian cities, 1951

| Region | Central place | Transpor- tation | Manufacturing | | | Extraction | Total |
			I	II	Total		
Hinterland							
Ontario	0	1	1	1	2	2	5
Other	7	4	5	12	17	3	31
Heartland							
Ontario	1	0	16	11	27	0	28
Quebec	0	0	14	1	15	1	16
Total	8	5	36	25	61	6	80

NOTE   The study includes all Canadian cities with a population of 10,000 or higher in 1951. The heartland is southern Ontario and Quebec. The five hinterland cities in Ontario are North Bay, Sault Ste Marie, Sudbury, Thunder Bay, and Timmins. Central place functions are retail trade, community service and government service.

more than fifty per cent of the excess employment is in manufacturing, and manufacturing II, in which manufacturing still dominates but is below 50 per cent (Table 3.2).

A marked contrast is evident between the heartland and hinterland manufacturing I cities. The thirty cities in the heartland group form highly integrated manufacturing regions with high degrees of industrial linkages and a range of industrial activity from blast furnaces to electronics and food processing. The hinterland group have isolated locations usually dominated by a single resource-oriented industry. Maxwell noted that the heartland–hinterland contrast is less marked in manufacturing II cities which tend to have important service and transportation functions.

Maxwell's study may be elaborated using 1961 county data to identify a four-class hierarchy of manufacturing activity and centres (also see chapter 4). The hierarchy is based on the incidence, by counties, for sixty-six industries (a) for all ninety counties in Canada with urban centres over 10,000, (b) for counties with a manufacturing I city in 1951, and (c) for counties with manufacturing II cities in 1951. Industries occurring in less than one-third of the three sets of counties are defined as sporadic. Industries occurring in two-thirds or more of each of the three sets of counties are classified as ubiquitous. The remaining industries are classified by their incidences on Maxwell's manufacturing I and II cities, with a residual group in which difference between the two incidences is very small. The proportions of each of these four industry-groups occurring in each Ontario county with a large urban centre produces a hierarchy of manufacturing centres from sporadic to manufacturing I, manufacturing II, and ubiquitous (Table 3.3). Two counties, York (Toronto) and

Table 3.3 A manufacturing hierarchy of Ontario urban counties (Tabulations specially prepared by Jeffrey P. Osleeb, from DBS (1960, 1966) and Maxwell (1965))

County	Largest city	1961 industry incidences					Maxwell 1951 classification
		Sporadic	Man. I	Man. II	Ubiquitous	Mean	
Metropolitan manufacturing centres							
York	Toronto	87.5	95.0	100.0	100.0	93.3	I
Wentworth	Hamilton	58.3	90.0	81.8	100.0	78.7	I
Manufacturing I centres							
Waterloo	Kitchener	50.0	80.0	45.5	100.0	68.0	I
Brant	Brantford	33.3	85.0	36.7	90.0	60.0	I
Middlesex	London	16.7	80.0	63.6	100.0	57.3	II
Wellington	Guelph	25.0	80.0	45.5	100.0	56.0	I
Peel	Brampton	45.8	60.0	45.5	90.0	56.0	*
Welland	Welland	33.3	70.0	27.3	90.0	53.3	I
Essex	Windsor	20.8	65.0	36.4	100.0	50.7	I
Lincoln	St Catharines	8.3	70.0	36.4	90.0	45.3	I
Carleton	Ottawa	16.7	60.0	45.5	100.0	44.0	Service
Oxford	Woodstock	12.5	55.0	45.5	90.0	44.0	I
Ontario	Oshawa	12.5	70.0	45.5	100.0	44.0	I
Perth	Stratford	16.7	50.0	18.2	90.0	41.3	I
Kent	Chatham	16.7	40.0	27.7	90.0	40.0	I
Northumberland	Cobourg	8.3	40.0	36.4	100.0	34.7	*
Peterborough	Peterborough	8.3	35.0	27.3	90.0	30.7	I
Stormont	Cornwall	4.1	35.0	0.0	70.0	26.7	I
Manufacturing II centres							
Halton	Burlington	20.8	55.0	72.7	100.0	53.3	*
Simcoe	Barrie	20.8	50.0	54.5	90.0	45.3	II
Hastings	Belleville	33.3	40.0	45.5	90.0	45.4	II
Lambton	Sarnia	4.2	30.0	36.4	100.0	29.3	I
Frontenac	Kingston	0.0	20.0	45.5	90.0	26.7	II
Thunder Bay	Thunder Bay	8.3	10.0	36.4	100.0	25.3	II

Table 3.3 Continued

County	Largest City	1961 industry incidences					Maxwell 1951 classification
		Sporadic	Man. I	Man. II	Ubiquitous	Mean	
Ubiquitous-industry centres							
Elgin	St Thomas	4.2	15.0	27.3	90.0	29.3	II
Grey	Owen Sound	0.0	25.0	27.3	100.0	29.3	II
Renfrew	Pembroke	12.5	25.0	18.2	70.0	28.0	II
Leeds	Brockville	3.5	28.6	30.7	80.0	26.7	II
Sudbury	Sudbury	4.1	20.0	9.1	100.0	22.7	Extraction
Nipissing	North Bay	0.0	10.0	9.1	80.0	17.3	Transportation
Cochrane	Timmins	0.0	0.0	9.1	100.0	16.0	Extraction
Kenora	Kenora	0.0	5.0	9.1	80.0	16.0	*
Algoma	Sault Ste Marie	4.2	5.0	9.1	80.0	16.0	I

*Cities below 10,000 in 1951 were not classified by Maxwell (see Table 2).

Industry definitions by SIC (Standard Industrial Classification of the Dominion Bureau of Statistics, 1960):
sporadic incidence below 33.3 per cent on all three sets of counties. Sporadic SIC groups are 107, 125, 133, 135, 143, 161, 163, 193, 211, 212, 213, 223, 245, 264, 268, 272, 292, 295, 323, 331, 332, 334, 338, 341.
manufacturing I incidence between 66.7 and 33.3 per cent, and higher on Maxwell's manufacturing I cities than manufacturing II. SIC groups are 131, 179, 183, 201, 216, 218, 231, 239, 244, 248, 273, 287, 291, 294, 297, 305, 353, 372, 375, and 385.
manufacturing II incidence between 66.7 and 33.3 per cent, and higher on Maxwell's manufacturing II cities than manufacturing I. SIC groups are 111, 128, 172, 174, 175, 221, 246, 247, 256, 271, 296, and 365.
ubiquitous incidence greater than 66.6 per cent on all three sets of counties. SIC groups are 101, 105, 129, 141, 261, 286, 288, 308, 348 and 381.

County definitions 1961:
metropolitan manufacturing centres more than 50.0 per cent of the sporadic industries.
manufacturing I centres 1961 50.0 per cent or less of the sporadic, 33.3 per cent or more of the manufacturing I industries, and a higher incidence on the manufacturing I industries than the manufacturing II.
manufacturing II centres 1961 50.0 per cent or less of the sporadic industries, 33.3 per cent or more of the manufacturing II industries, and a higher incidence on the manufacturing II industries than the manufacturing I.
ubiquitous-industry centres 50.0 per cent or less of the sporadic and less than 33.3 per cent of the manufacturing I and II industries.

Wentworth (Hamilton), have such high proportions of the sporadic industries that they are classed as metropolitan manufacturing centres. Both were ranked as manufacturing I for 1951 by Maxwell. The other 1951 manufacturing I cities retain this classification for 1961 with two exceptions, Sarnia, relegated to manufacturing II and Sault Ste Marie which ranks only as ubiquitous; in addition, the manufacturing I counties include Middlesex (London) classed as manufacturing II in 1951, Ottawa, whose dominant function is government service, and a number of centres that were below 10,000 in 1951.

The 1961 classification identifies a continuous belt of metropolitan and manufacturing I centres extending from Oshawa, around Lake Erie, to Welland (the Golden Horseshoe) and, with the reclassification of London as manufacturing I, across southwestern Ontario to Sarnia. Thirteen of southwestern Ontario's twenty counties are incorporated in this belt, including the top ten on the mean incidence index. By contrast, of the thirty-four counties in the remainder of Ontario, only five counties rank as manufacturing I centres. These five are all in eastern Ontario, but even if manufacturing II and ubiquitous industry counties there are also considered, industrial counties in eastern Ontario comprise less than half the total, and are scattered. These results for Southern Ontario concur with Maxwell's 1951 findings. The major difference is the emergence in the 1961 analysis of the Kenora–Nipissing belt of Northern Ontario counties with ubiquitous-type industries. The manufacturing hierarchy of Ontario's large urban centres thus displays broad, regional patterns, with metropolitan manufacturing I type counties predominant in southwestern Ontario, manufacturing I and II in the Eastern Ontario–Georgian Bay area, and ubiquitous industry in Northern Ontario.

Central Place Hierarchy and Functional Regions
In one of the most comprehensive analyses to date of the Ontario urban hierarchy and functional regions in the central-place framework, Hans Carol designates urban regions primarily by professional and medical services, and shopping and urban recreation travel (Carol 1969). (The subject is also discussed in chapter 4.) Carol recognizes three orders: the highest, comprising Toronto; four high-order centres, Ottawa, Hamilton, London, and Windsor; and fifteen middle-order centres. Carol's study is a significant contribution to understanding the centripetal organization of space, but Gertler has noted the functional change of metropolitan areas from being only centripetal labour markets to serving also as centrifugal leisure markets (Gertler 1969). One measure of centrifugal leisure forces is the city locations of cottage owners (Wolfe 1968; Dean 1969). A com-

parison of the regions of cottage ownership with Carol's functional regions for highest and high-order centres suggests a competitive displacement of the central-place regions which underlines the importance of sheer urban size and geographic position in delimiting urban regions for space-consuming activities.

Urban Growth Complexes

Recognizing the importance of the size of urban population and the deficiencies of rigid areal definitions of urban areas, 'urban growth complexes' have been defined for Ontario and Quebec encompassing urban areas of metropolitan stature, in terms of population size and diversity of infrastructure, within which location decisions are largely unaffected by regional considerations (Bourne & Baker 1968). The largest of these is an extended Toronto–Hamilton complex called Lake Ontario, with a population of just over three million, making it a little larger than Montreal. The next largest in Ontario is Ottawa–Hull with over half a million; Kitchener, London, Windsor, and Niagara Falls, all in southwestern Ontario, form a third group of quarter-million population urban growth complexes.

A distinctive characteristic of Canadian metropolitan centres is the proportion of their population that is foreign born. In 1961, 15.6 per cent of Canada's population was foreign born, compared with 5.4 per cent in the United States. The proportion of foreign born, especially United Kingdom born, is highest for the Lake Ontario urban growth complex, and the proportion of post-World War II immigration to this area, and to southwestern Ontario generally, is even higher. In general, immigration rates to Ontario metropolitan centres are related to their population size (Table 3.4). Metropolitan growth has also been bolstered by in-migration, so that the population growth of metropolitan areas in the intercensal period, 1951–61, was 44.8 per cent compared with 20.3 in non-metropolitan areas. Urban growth has tended to be related to both heartland–hinterland and urban hierarchical forces (see chapter 4).

THE INTER-METROPOLITAN AXIS PATTERN

Interaction

Metropolitan centres are linked by an axis along which the rate of commodity flows, interaction, and development tend to be proportional to the size of centres and inversely proportional to their distance apart. A political or physical boundary between two metropolitan centres will tend to reduce the level of interaction between them, although a spatial mo-

Table 3.4   The percentage of foreign-born population in Ontario metropolitan areas, 1961 (Kasahara 1963)

Metropolitan area	Population total	1961 rank	Foreign born as a % of total population	
			%	Rank
Toronto	1,824,481	1	33.3	1
Hamilton	395,189	2	28.0	2
Windsor	193,365	3	23.1	3
London	181,283	4	21.1	4
Kitchener	154,864	5	20.4	5
Sudbury	110,694	6	16.1	6

NOTE   The figure for Ottawa could not be calculated as the metropolitan area extends across the Ontario–Quebec boundary. Population rank and rank by per cent of population foreign born are perfectly correlated.

mentum factor, which is proportional in strength to the distance of the centres from the boundary, appears to operate (Mackay 1958; Ray 1965; Wolfe 1968). A metropolitan centre located between two centres may reduce interaction between them by serving as an intervening opportunity, although this effect also may diminish as the distance between the two centres increases.

Air Line Traffic

Little data on interurban interaction exists, with the important exceptions of airline passenger traffic and telephone calls (see also chapter 4). The volume of passengers flying between pairs of Canadian cities is related to the size of the cities and their distance apart (Kerr 1968). Toronto is the major domestic-passenger node in Canada and passenger flows between Toronto and other large Canadian cities, particularly Vancouver and Winnipeg, are greater than are predicted from their size and distance. The data also suggest that Toronto acts as an intervening opportunity between the western periphery and Montreal. For example, air passenger traffic between Vancouver and Montreal in 1964 numbered only 26,000 compared with 60,000 between Vancouver and Toronto.

The air routes on which Canadian carriers transported over a hundred thousand passengers in 1964 were, in order of importance, Toronto–Montreal, Toronto–New York, Montreal–New York, and Toronto–Ottawa. The complete data on passenger traffic reveal the importance of the Toronto–Montreal axis and its links with both the Canadian hinterland, particularly Vancouver and Winnipeg, and the United States heartland, particularly New York and Chicago (Wolfe 1968). It also provides an indication of the urban hierarchy and the way in which the metropolitan centres link the national heartland and hinterland.

United States Subsidiaries

The number of subsidiaries in a Canadian city that is controlled by corporations within a United States metropolitan area also follows the interaction model and is proportional to the number of manufacturing establishments in that metropolitan area, and inversely proportional to distance from that metropolitan area (Fig. 3.7). New York controls 307 Canadian subsidiaries, whereas Boston, much smaller but a little closer to Canada, has 48. Chicago and Los Angeles each have about the same number of manufacturing establishments, yet Chicago controls 197 Canadian subsidiaries compared with the more distant Los Angeles which controls only 45.

Because of the contribution of United States subsidiaries to the manufacturing industry in Canada, there are two important corollaries to the interactance model: first, regional economic development and urban growth in Canada will tend to reflect the economic health of adjacent regions of the United States, and second, the Canadian regions most likely to attract a large number of United States subsidiaries are those such as southwestern Ontario that are closest to the American manufacturing belt.

Two additional elements in the location of United States subsidiaries occur. The stronger of these is the tendency of subsidiaries to locate in the geographic sector that links the parent company with Toronto, the point of highest market potential. Toronto provides the optimal market location for American subsidiaries and few parent companies locate subsidiaries beyond it. Industrial interactance between a Canadian city and a United States city is severely restricted wherever Toronto becomes an intervening opportunity, eclipsing sectoral affinity.

Furthermore, the distance that a parent company penetrates into Canada to locate a branch plant is directly proportional to the distance of the parent company from Canada. The Detroit manufacturer, for instance, can evade the Canadian tariff barrier and the prejudice against foreign products by locating a branch plant across the Detroit River. The marginal benefits of locating closer to the centre of the Canadian market may not compensate for losing the convenience of operating the subsidiary close to the parent company. Consequently, 34 of Detroit's 87 Canadian subsidiaries are located in Windsor where they comprise more than half the total of United States subsidiaries. Eight of Seattle's 11 Canadian subsidiaries are in Vancouver. Detroit controls only 20 subsidiaries in Toronto, and Seattle, none. By contrast, Los Angeles has half of its

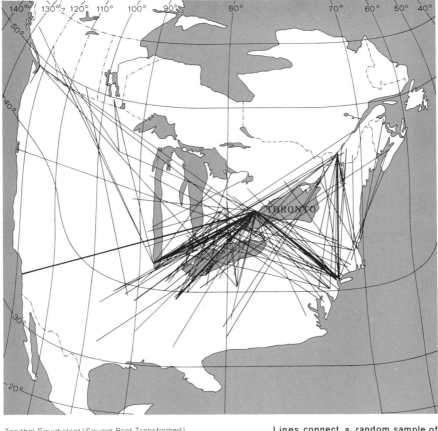

Zenithal Equidistant (Square Root Transformed)
Origin Toronto

Toronto   100   300   600

Lines connect a random sample of
170 United States subsidiaries in
Canada to their parent companies

Number of Subsidiaries
1962

1   2-3   4-6   7-10 11-15 16-20

3.7

United States - Canada, Industrial Interactance

subsidiaries in Toronto, but none in Windsor; in general, the proportion of an American metropolitan area's Canadian subsidiaries located in Toronto increases with increasing distance of the metropolitan area from the Canadian border.

The rapid industrial and urban growth of southwestern Ontario, and the contrast it presents with that of eastern Ontario may be explained to a large extent by the impact of the United States on Canadian development, and by the concentration of this impact along development axes

linking Toronto, across southwestern Ontario, to the United States manufacturing belt.

Urban growth can be expected to concentrate along these development
axes so that one geographer has called them 'urban corridors' and de-
scribes them as the main streets for the Ontario megalopolis of the twenty-
first century (Whebell 1969). The location of modern highways, notably
the Queen Elizabeth and the MacDonald-Cartier Highways, is both a
response to the traffic generated along these corridors and an inducement
to the location of new industry and accelerating urban development.

SUMMARY

The spatial structure of the Ontario economy is a product of the centripetal
and centrifugal forces that have operated during its growth. The spatial
components identified are heartland–hinterland, urban hierarchy, and
development axis. These components were early creations of the European
occupance. The changing economic geography of Ontario thus appears to
reflect the changing relative prominence of these basic components rather
than the emergence of new ones. Much more attention needs to be given
to the allometry of economic development, and the changing form of the
space economy with increasing size. The evidence available suggests that
centripetal forces are strengthening at the national level. They may be
strengthening also at the regional level. It therefore seems probable that
Toronto and southwestern Ontario, located within the Canadian heartland
and comprising the land links to the United States heartland, will gain an
increasing share of Ontario's and of the nation's population and economic
activity. The forces promoting disproportionate growth within the Ontario
and national economy are unlikely to be reversed or halted as a result of
better understanding of their nature and more effective regional planning;
but better understanding and planning will increase both the chances of
reducing the associated problems of regional economic disparities and the
hopes of improving the relationship between man and his physical environment.

4 The Urban Network

J.U. MARSHALL

Ontario today is the most urbanized of the major regions of Canada (Table 4.1). Seventy per cent of its population lives in cities of 10,000 or more. With respect to the proportion of the regional population in cities of 10,000 or more, Quebec and British Columbia are close behind Ontario. The Prairie Provinces occupy an intermediate position, while the Atlantic Provinces lag far behind. The North, comprising the Yukon and Northwest Territories, might be considered a special case; no town reaches a population of 10,000, but a high proportion of the region's population lives in small nucleated settlements, and the conventional distinction between 'urban' and 'rural' is perhaps inappropriate.

Table 4.1 also indicates that Ontario is the only part of Canada in which centres of metropolitan size (here defined as 100,000 or more) are particularly concentrated. In Quebec, only Montreal, Quebec City, and Chicoutimi–Jonquière reach this threshold, while in British Columbia there are only two metropolitan centres, Vancouver and Victoria. By contrast, Ontario contains nine metropolitan centres, with the result that the largest centre, Toronto, accounts for a much lower share of the total provincial population than is the case in Quebec and British Columbia.

The smallest centres considered in Table 4.1 are those of 10,000 population, but it may be argued that many places much smaller than this are urban in character. Accordingly, Table 4.2 presents selected data for all nucleated places in Ontario which had a population of at least 1000 at the time of the 1966 census, including unincorporated as well as incorporated centres. In Table 4.2 municipalities have been combined where appropriate in order to obtain the closest possible match with discrete built-up localities as they exist on the ground. In some cases, including that of Toronto, this has involved the creation of combinations of municipalities different from those recognized for metropolitan centres in the census.[1]

1 The Census Metropolitan Areas of Toronto, Hamilton, and Kitchener are

Table 4.1 Levels of urbanization in Canada's major regions (calculated from the *1966 Census of Canada*)

Region	Per cent of regional population in cities of 10,000 of more	Per cent of regional population in the largest city	Number of cities of 100,000 or more
Ontario	70.2	28.6	9
Quebec	66.4	42.2	3
British Columbia	65.3	47.6	2
Prairie Provinces	51.5	15.0	5
Atlantic Provinces	37.0	10.0	4
The North	(No centre of 10,000 or more population)		

Table 4.2 reveals a clear distinction between the northern and southern parts of the province. Southern Ontario, comprising all territory south of the Lake Nipissing lowland, contains 81 per cent of the province's 262 urban centres. The significance of this figure becomes apparent when it is realized that Southern Ontario contains a mere 12 per cent of the province's total area. On the average, Southern Ontario has one urban centre for every 192 square miles of land area, while in Northern Ontario the ratio is one centre for every 5,954 square miles. The relative emptiness of Northern Ontario is one of the fundamental features of the province's spatial structure, and is attributable to the environmental limitations described in the opening chapter of this volume. A short growing season combines with the limited arable land of the Canadian Shield to keep agriculture to a minimum, and the area's low market potential inhibits the development of diversified manufacturing and central place towns. With two exceptions, the larger centres of Northern Ontario are sustained primarily by mining and the pulp and paper industry. The exceptions are Thunder Bay, which thrives on the trans-shipment of Prairie grain bound for eastern markets, and Sault Ste Marie, which ranks behind Hamilton as Canada's second-largest steel-milling centre.

In the sections which follow, three selected aspects of the Ontario urban network are described: recent population change, with special reference to the relative significance of the natural and migrational components of change; variations in employment profiles; and nodal structure as reflected by newspaper circulation and flows of telephone calls. These topics have

seriously overbounded. As a result of the necessary corrections, the 1966 population given for Toronto in Table 4.2 is smaller than the 2,158,496 appearing in the census.

Table 4.2 The Ontario urban network (calculated from the *1966 Census of Canada*)

Type of place	Number in Southern Ontario	Number in Northern Ontario	Total number of centres	Per cent of all centres	Total population (1966)	Per cent of 1966 provincial population	Per cent of 1961 provincial population	Number of centres declining (1961–6)
Toronto	1	—	1	0.4	1,989,542	28.6	27.2	—
100,000–500,000	7	1	8	3.1	1,636,753	23.5	22.9	—
30,000–99,999	13	4	17	6.5	911,568	13.1	12.4	—
10,000–29,999	18	4	22	8.4	350,799	5.0	5.1	3
5000–9999	29	7	36	13.7	257,145	3.7	3.9	6
1000–4999	143	35	178	67.9	391,695	5.6	5.8	36
Total urban	211	51	262	100.0	5,537,502	79.5	77.3	45
Total rural					1,423,368	20.5	22.7	
Total Ontario					6,960,870	100.0	100.0	

been selected in order to achieve, in a limited space, a view of the spatial patterns of the network's most basic demographic and economic characteristics.

Recent Population Change

Figure 4.1 shows the pattern of population change in the Ontario urban network for the period 1961–6. Positive rates of change are mapped according to a geometric scale which is based on the overall provincial growth rate of 11.62 per cent for the same period. For clarity, Figure 4.1 excludes centres below 5000 in population. However, an examination of the population changes of these small places shows that their inclusion in Figure 4.1 would simply emphasize the principal regional variations which already appear.[2]

A salient feature of the pattern of recent growth is the strong concentration of rapidly growing centres in the crescent of land around the western end of Lake Ontario from Oshawa to Kitchener to Niagara Falls, an area whose urban growth is described in chapter 2. Within this crescent, almost every centre is growing faster than Ontario as a whole. The portion immediately adjacent to the lake is known as the 'Golden Horseshoe,' but Figure 4.1, as well as Figures 2.3 and 2.4 in chapter 2, suggest that the area denoted by this popular term should be expanded westwards to include Kitchener and other rapidly growing centres of the middle Grand River system. The Golden Horseshoe, so expanded, contains almost 60 per cent of the 5.5 million people in Ontario who live in urban centres of at least 1000 population.

In a broader regional context, the Golden Horseshoe may be viewed as the principal node in a system of transportation corridors which for almost two centuries has served as the locus of population concentration and economic development in Southern Ontario (Whebell 1969). The main axis of this transportation system runs from Montreal via the St Lawrence Valley and the north shore of Lake Ontario to Toronto, thence by way of Kitchener to the Thames Valley and London, and finally due west to the international border at Sarnia. This axis may be termed the Grand Trunk corridor after the early railroad of the same name. Four branch corridors emanate from this central axis: the Ottawa Valley corridor; the Sagamo corridor, which runs north from Toronto to Lake Simcoe and ultimately Northern Ontario; the Niagara corridor, running from Hamilton to Buffalo and included within the modern Golden Horseshoe; and the Long Woods

2   In dealing with the five-year period 1961–6, Figure 4.1 complements the map of population change for the decade 1951–61 appearing as Plate 29 of the *Economic Atlas of Ontario* (Dean 1969).

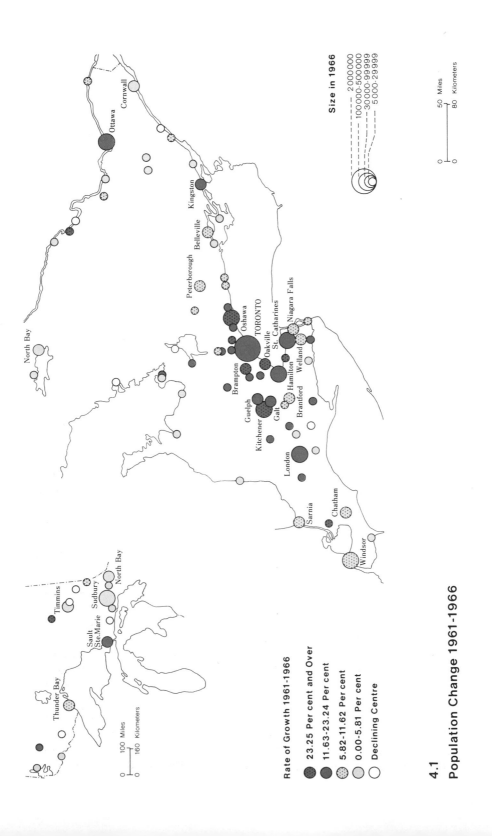

Rate of Growth 1961-1966

23.25 Per cent and Over

11.63-23.24 Per cent

5.82-11.62 Per cent

0.00-5.81 Per cent

Declining Centre

Size in 1966

2 000 000
100 000-500 000
100 000-99 999
30 000-99 999
5 000-29 999

Thunder Bay
Timmins
Sault Ste.Marie
Sudbury
North Bay

North Bay
Ottawa
Cornwall
Kingston
Belleville
Peterborough
Oshawa
TORONTO
Oakville
St. Catharines
Niagara Falls
Welland
Hamilton
Brantford
Galt
Kitchener
Guelph
Brampton
London
Sarnia
Chatham
Windsor

4.1

Population Change 1961-1966

corridor, which follows the Thames Valley below London to Chatham, and then connects with the international border at Detroit. The Ottawa Valley and Sagamo corridors connect the Grand Trunk route with the forests and mines of Northern Ontario, while the Niagara and Long Woods corridors serve as links with the main manufacturing belt of the United States to the south.

Generally speaking, the cities of Southern Ontario which are experiencing the most rapid growth are located on, or very close to, these principal transportation corridors. Ottawa, Kingston, Barrie, and London are good examples drawn from outside the Golden Horseshoe core area. It does not follow, however, that slow-growth centres are necessarily remote from the main travelled routes. Owen Sound and Goderich are indeed off the beaten track, but Trenton and Prescott lie right on the Grand Trunk corridor. One may speculate that the same process of centralization which, at the provincial scale, has created the Golden Horseshoe, operates also at a more local scale along the corridors radiating from that nexus. The larger corridor centres are pulling farther ahead, while their smaller neighbours have difficulty holding their own. For example, Kingston advances while Belleville and Brockville grow only slowly; and Prescott, which lost an important trans-shipment function with the opening of the St Lawrence Seaway in 1959, actually declines.

In Northern Ontario, most towns have grown up where economically workable minerals occur or where conditions are suitable for the establishment of pulp and paper mills. On the whole, rates of urban growth in Northern Ontario are markedly lower than those in the south. Rapid growth (other than the apparent rapid growth resulting from territorial annexation, as at Kapuskasing) normally occurs only with an expansion in the pulp and paper industry or with the opening of a new mine. Mining centres are always liable to decline, not merely from exhaustion of the resource, but also as a result of competition from new sources. The decline of the uranium centre of Elliot Lake, however, is a special case, for here the crucial factor was the decision of the United States, the major customer, to cease stockpiling uranium.

For Ontario as a whole there is no significant relationship between size of centre and rate of growth. Taking the 84 places of 5000 and more population in 1966, the Spearman coefficient of rank correlation between size in 1961 and rate of growth 1961–6 is +0.204, which is not significant at the 0.05 confidence level. Two of the factors accounting for this low level of association are worth mentioning. First, a small centre is just as likely as a large centre to experience a high rate of growth owing to territorial annexation. Oakville (large) and Kapuskasing (small) are cases in point. Secondly, while the largest cities generally do have fairly high growth rates (the northern centres of Sudbury and Thunder Bay being exceptions), equally high rates are found among small centres within commuting distance of a metropolis. This tendency is particularly evident in the vicinity of Toronto, with Ajax, Markham, Richmond Hill, Aurora, and Orangeville all among the top 15 places in Ontario when ranked by rate of growth.

In addition to inspecting overall changes in population, it is of interest to consider the relative significance of the natural and migrational components of these changes. Provided a locality retains a fixed administrative boundary during the period of study, data on births and deaths may be used to obtain a measure of the natural component of that place's population change. The difference between this natural component and the overall change in the place's population then gives a measure of the migrational component of change. Places may then be classified according to the directions and magnitudes of the two components of change.

An example will serve to clarify the technique. The example relates to the Toronto metropolitan area, but the method can be applied to any administrative area for which the relevant data are available. The total difference between the 1961 and 1966 populations of Toronto as recorded in the censuses is 295,880. During the same five-year period, according to the statistics compiled by the provincial registrar general, births exceeded deaths in Toronto by 140,703. Subtracting this natural component of change from the overall change reveals the fact that Toronto experienced a net in-migration of 155,177 persons. Toronto thus gained population both by natural increase and by net in-migration, with the latter being slightly the larger of the two components.[3]

The inset on Figure 4.2 shows how centres may be classified into eight types on the basis of the relative significance of the natural and migrational components of change (Webb 1963). The vertical axis represents natural change, while the horizontal axis represents migrational change; where the two axes intersect, both have a value of zero. Each of the four quadrants is divided into two sectors on the basis of the relative magnitudes of

3   Comparability between the federal censuses and the provincial vital statistics is hampered to a small degree by the fact that the latter relate to calendar years while the census is usually taken in late spring. This problem is circumvented by halving the total numbers of births and deaths for 1961 and 1966, and adding the half-totals for these end-years to the full totals for 1962 through 1965. The result is a close approximation to the five-year intercensal period. The data on births and deaths are available in the annual *Vital Statistics Report* published by the office of the Ontario Registrar General.

the two components of change. Figure 4.2 shows the map of these eight population types (A through H) for 88 places of 2500 and over for which the necessary data are available.[4]

The most spatially concentrated type of centre in Figure 4.2 is type C. About three-quarters of the type C centres appearing on the map are clustered in the Golden Horseshoe, particularly in the part of this region closest to Toronto. This indicates that the Toronto metropolitan community, stretching from Oshawa in the east to Brampton and Oakville in the west, is the mecca of migrants from elsewhere in Ontario and also from outside the province, including arrivals from overseas. Outside the Golden Horseshoe, the only centre of 5000 or more to achieve type C status is Orangeville. This old market town lies today within commuting distance of the Golden Horseshoe, and is increasingly taking on the character of a dormitory satellite.

There is some tendency for centres of types A, B, and H to form concentric belts around the cluster of type C towns centred on Toronto. This demographic succession is particularly evident to the north and northwest, where the influence of Toronto as a population magnet is least affected by other centres of attraction. Moving north from Toronto, one first encounters the type B centres of Richmond Hill and Aurora. These towns experience both natural increase and net in-migration, but the former predominates, since migrants are attracted less to these centres than to the type C places closer to Lake Ontario. Farther out, in a zone approximately 60 to 120 road miles from Toronto, one finds type A centres; these still enjoy overall growth as a result of normal natural increase, but net migration now effects a loss rather than a gain. Collingwood, Owen Sound, and Penetanguishene are representatives of this zone. Still farther out, net out-migration more than offsets natural increase, and centres of type H, undergoing absolute decline, are found. Parry Sound and Kincardine are examples. This zonation of population types outwards from Toronto is by no means sharply defined, but the map nevertheless provides circumstantial evidence of a general process whereby the cities of the Golden Horseshoe are augmenting their populations at the expense of outlying centres characterized by chronic out-migration. Recent work employing a cohort survival technique indicates that migrants in Ontario, as else-

4   Data on births and deaths are not separately available for centres which are unincorporated, and the method of analysis is inapplicable if a centre's boundary (a) understates the extent of the built-up area or (b) changed during the period of study. In order to indicate where the chief gaps in the analysis lie, Figure 4.2 shows centres of 30,000 and over which, for one reason or another, had to be disregarded.

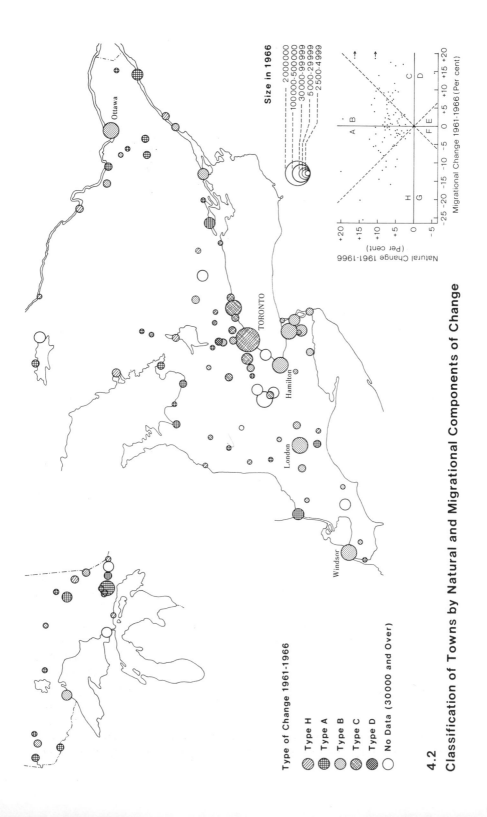

Size in 1966

2 000 000
100 000-500 000
30 000-99 999
5 000-29 999
2 500-4 999

Migrational Change 1961-1966 (Per cent)

Natural Change 1961-1966 (Per cent)

Ottawa

TORONTO

Hamilton

London

Windsor

Type of Change 1961-1966

Type H
Type A
Type B
Type C
Type D
No Data (30000 and Over)

4.2   Classification of Towns by Natural and Migrational Components of Change where, are chiefly in the 15–34 age group (Whebell 1968), and it may be surmised that the movements implied by Figure 4.2 are motivated primarily by people's perception of economic opportunity.

Northern Ontario contains no centre of type C, but more than its share of centres in types A and H. Net out-migration is a fact of life for all but four of the centres in Northern Ontario on Figure 4.2. It may be inferred that the resources of the north cannot provide enough jobs to absorb the region's natural increase in population, and that Northern Ontario as a whole acts as a source of migrants for the more economically diversified cities of Southern Ontario and elsewhere.

Taking the 88 cities on Figure 4.2 as a whole, overall rates of population change are associated much more closely with migrational rates of change than with natural rates of change. The Spearman coefficient of rank correlation between natural rate of change and overall rate of change is +0.295, which is barely significant at the one per cent confidence level. By contrast, the rank correlation between overall rate of change and migrational rate of change is +0.895, a very highly significant result. There is no doubt that overall rates of change in the Ontario urban network are primarily a function of the direction and local relative magnitude of the migrational component of population change.

Variations in Employment Profiles

Within a developed area such as Ontario, it is clear that the most ubiquitous city-sustaining economic activities are manufacturing and the regional service function. This assertion is supported both by the general literature on urban economics (Duncan *et al.* 1960; Thompson 1965), and by numerous 'functional classification' studies in the geographic literature (including Nelson 1955; Alexandersson 1956; Smith 1965; and Maxwell 1965). Given the present limitations of space, it is proposed to exploit the importance of manufacturing and the regional service function in order to produce a simple yet informative typology of Ontario cities. Compared to the work of Maxwell (1965) at the national scale, the analysis presented here is methodologically less sophisticated, but employs data of more recent date (1961 instead of 1951), and includes a larger number of Ontario cities, the minimum population being 5,000 instead of 10,000.

The method may briefly be described as follows. First, 'manufacturing' is taken to mean the labour force included in the census category of the same name, while 'regional service' is defined as a combination of the census categories of Retail Trade and Wholesale Trade. For each city, the percentages of the labour force in manufacturing and in regional service are calculated, and these two sets of percentages are each standardized to

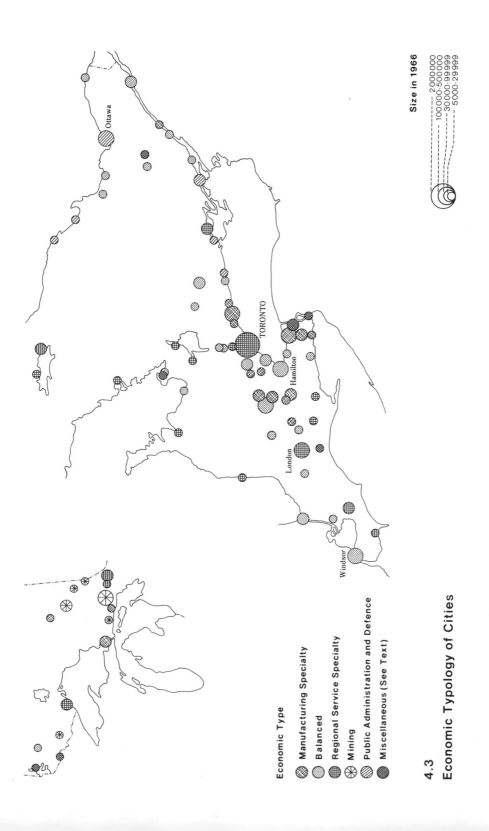

Economic Type

Manufacturing Specialty

Balanced

Regional Service Specialty

Mining

Public Administration and Defence

Miscellaneous (See Text)

Size in 1966

2 000 000
100000-500000
30000-99999
5000-29999

Ottawa

TORONTO

Hamilton

London

Windsor

4.3
Economic Typology of Cities

zero mean and unit variance. Next, the set of cities is partitioned into two groups. The first group contains those cities in which both manufacturing and regional service show negative standard scores; neither of the two most ubiquitous activities reaches its mean percentage of the labour force. On the basis of an inspection of their complete labour force profiles, these cities are each subjectively allocated to one of three categories, namely Public Administration and Defence, Mining, and Miscellaneous.

In each city of the second group, at least one of the two key activities shows a positive score. In each city, the simple algebraic difference between the scores is calculated. If the absolute value of this difference is equal to or greater than unity (that is, the value of one standard deviation on the standardized scale), the city is regarded as specializing in the activity having its score farthest from the mean in the positive direction. For example, the scores for Leamington are +0.346 in manufacturing and +1.750 in regional service, with the result that Leamington is classed as Regional Service Specialty. For Welland, the scores are +1.511 in manufacturing and −1.166 in regional service, with the result that Welland is classed as Manufacturing Specialty. If the difference between the standard scores is less than 1.000, the city is classed as Balanced, since neither of the two key activities is dominant over the other. The critical value of 1.000 was established quite arbitrarily, but it happens that the three subgroups thus created are not greatly different in size.[5]

These procedures resulted in the typology mapped in Figure 4.3. As expected, centres classed as Manufacturing Specialty are strongly concentrated in the Golden Horseshoe, lending support to the idea that this portion of Ontario may be regarded as a Canadian extension of the American manufacturing belt. Thirteen of the 17 cities specialized in manufacturing are situated in the triangle bounded by Bowmanville, Woodstock, and Port Colborne. Of the four remaining centres in this class, three are found in Northern Ontario. All three are important pulp and paper centres, and Sault Ste Marie in addition is the site of a major steel mill. Outside the Golden Horseshoe and the north, the only centre specialized in manufacturing is the town of Prescott in eastern Ontario.

Compared with the manufacturing centres, cities in the category of Regional Service Specialty are spatially less concentrated, but nevertheless 13 out of 16 are found in Southern Ontario southwest of a line from

5 No attempt was made to separate the supposedly non-basic portion of each economic activity from its basic or 'export' portion (Alexander 1954). This decision was based upon the convincing arguments put forward by Pratt (1968) in his critique of the standard 'minimum requirements' technique designed by Ullman and Dacey (1960).

Belleville to Parry Sound. Except for Toronto and its satellite of Richmond Hill, centres specializing in regional service are conspicuously absent from the Golden Horseshoe. The appearance of Toronto as a regional service centre reflects its role as Ontario's primate metropolis, performing a wide variety of services for a hinterland which is difficult to define but which includes all of Ontario except the extreme east (oriented to Montreal), and the far northwest (oriented to Winnipeg). Toronto's status as a regional service centre should not be taken to imply that it lacks importance in manufacturing. In fact, measured in absolute terms, Toronto has by far the largest manufacturing employment of the 79 cities analysed.

Table 4.3 shows the average employment profiles of the six classes of cities in the typology. The average profile for all cities taken together is also shown. In the case of the cities classed as Manufacturing Specialty, manufacturing on the average employs just under one half of the labour force, with all other activities falling below their provincial means. Manufacturing remains above average in the Balanced cities, but is joined there by both retailing and financial activities. In the cities classed as Regional Service Specialty, manufacturing drops below its provincial mean, but these cities are above average in six of the nine categories of industry listed. As might be expected, the profile of the Balanced cities is most similar to the overall provincial profile.

For two of the three remaining types of city, the average employment profiles are quite unequivocal. Centres classed as Public Administration and Defence are well above average in this category of industry, and below average in all other categories. Similarly, centres classed as Mining are above average only in the census category of Extraction. All six of the mining centres are situated in Northern Ontario, where the ancient rocks of the Canadian Shield form one of the world's greatest storehouses of mineral wealth. The principal products are nickel and copper (Sudbury), iron (Sudbury and Atikokan), gold (Timmins and Kirkland Lake), silver (New Liskeard–Cobalt), and uranium (Elliot Lake).[6]

The six administrative cities form two groups of three, one group occupying the north shore of Lake Ontario (Cobourg, Trenton, Kingston), and the other the Ottawa Valley (Ottawa, Pembroke, Deep River).

6   In recent years zinc has risen to a prominent position among Ontario's mineral products; by 1967, zinc ranked fourth in value behind nickel, copper, and iron ore. Most of Ontario's zinc comes from the copper–zinc mines at Manitouwadge, a town of some 3000 located north of Lake Superior in the eastern part of the District of Thunder Bay. Manitouwadge does not appear on Figure 4.3 because the necessary labour force data are not available for unincorporated centres or for centres with fewer than 5000 population.

Table 4.3   Mean employment profiles by type of city (calculated from the *1961 Census of Canada*)

	Type of city[a]						
Industry	I (%)	II (%)	III (%)	IV (%)	V[b] (%)	VI (%)	All cities
Extraction	0.5	0.3	0.7	0.1	40.2	0.3	1.5
Manufacturing	49.4	41.1	26.7	18.1	9.7	28.1	36.0
Transportation, Communication, and Utilities	7.5	8.7	12.1	6.3	7.3	17.0	9.7
Retail Trade	12.5	14.9	16.4	11.5	13.0	13.6	14.2
Wholesale Trade	3.1	3.6	5.1	2.6	3.3	2.5	3.6
Finance, Insurance, and Real Estate	2.7	3.5	4.4	3.1	2.7	2.7	3.4
Personal Services	6.7	8.0	8.9	7.2	7.4	9.4	8.0
Community and Business Services	13.2	14.8	17.8	15.4	12.2	20.2	15.5
Public Administration and Defence	4.5	5.2	8.1	35.7	4.2	6.3	8.2

Columns may not sum to 100.0 owing to rounding

a   City types: I Manufacturing Specialty; II Balanced; III Regional Service Specialty; IV Public Administration and Defence; V Mining; VI Miscellaneous.
b   Figures in this column represent only Sudbury and Timmins. The other four centres in the Mining class are unincorporated, and labour force data are not available.

Kingston, originally an important garrison town commanding the eastern entrance to Lake Ontario, is today the home of the Royal Military College and the site of major federal prisons. Farther west, Trenton is the location of one of Canada's largest military air bases, while Cobourg contains a major military supply depot (the latter to be closed in 1971). In the Ottawa Valley, Deep River is the principal dormitory settlement for the nearby atomic energy research installation at Chalk River. Pembroke also houses some of the Chalk River staff, but in addition is a dormitory for the large army base of Camp Petawawa. Ottawa, the national capital, completes the list of administrative centres.

As Table 4.3 shows, centres classed as Miscellaneous are above the provincial averages in Personal Services, Community and Business Services, and the somewhat complex category of Transportation, Communication, and Utilities. This latter category is dominated by employment in railroad yards, which are especially prominent at Kenora and Smiths Falls.

However, it also includes employment in the power industries, and the major hydroelectric installations at Niagara Falls are the main reason for the inclusion of this city in this class. Regarding Community and Business Services, the decisive element is the presence of major psychiatric hospitals in St Thomas, Smiths Falls, and Penetanguishene. Finally, the prominence of Personal Services results from the fact that this category includes employment in hotels, motels, and restaurants, all of which are abundantly represented in the tourist centres of Niagara Falls, Kenora, and Fort Frances.

Nodal Structure
No city stands isolated from the rest of the world. Cities are connected to one another, and to the countryside, by a wide variety of social and economic flows. These flows of people, goods, and messages across the land are extremely complex, yet within the complexity certain quasi-stable patterns resulting from repeated interaction can be discerned. Most notable, perhaps, is the pattern of city-centred nodal regions, each consisting of a market centre and a tributary area in symbiotic co-existence. This polarization of space operates at several scales, leading to the familiar concept of a nested hierarchy of central places and trade areas (Berry 1967). Although this concept emerged originally from a specific concern with the retailing and personal service sectors of the economy, (Christaller 1933), it is relevant in greater or less degree for other important types of activity, including wholesaling, the daily journey to work, rural–urban migration, and the recreational behaviour of city residents.

In Figure 4.4, the larger cities of Southern Ontario are classified as major and minor shopping centres, with Toronto in a class by itself as the province's primate metropolis. (Northern Ontario is excluded because the concept of tributary area has limited application where there is no continuous rural population.) The classification is based partly upon population size and dollar volume of retail sales, but also upon the extent to which pairs of centres are mutually independent in the sense that the residents of one do not normally shop in the other. A good example of the application of this latter criterion concerns the cities near the western end of Lake Ontario. Where the shopping behaviour of the residents is concerned, Kitchener and Hamilton are independent of each other, though shoppers from both cities regularly visit the primate centre of Toronto. By contrast, Kitchener and Guelph are not mutually independent, since the residents of the latter habitually shop in the former. Hamilton and Kitchener are thus on a par as major shopping centres, while Guelph is classed as a minor shopping centre. The collection of flow data relating to this concept of mutual independence is part of a continuing research programme, but experience to date supports the classification as it appears on Figure 4.4.

Five cities on Figure 4.4 are identified as 'large centres with limited regional influence.' Owing to the proximity of larger, regionally dominant centres, these five cities effectively have no trade areas which they can claim to dominate. Niagara Falls and Welland are overshadowed by St Catharines, Oakville is largely a dormitory for Toronto, and Galt and St Thomas are little more than satellites of Kitchener and London respectively.

Although the classification is hierarchical in form, comparison with the locational models of central place theory is precluded by the small number of cities mapped and by the great variability of the employment profiles of these cities. An evaluation of the applicability of the Christallerian models would involve a rigorous assessment of the retail and service activities of centres of all sizes, together with careful testing of the degree of resemblance between the real pattern and the spatial arrangements derived from theory. Detailed work of this type has been completed northwest of Toronto in the territory served by the minor shopping centres of Barrie and Owen Sound. With allowance made for variations in the disposable income surface, and for the fact that not all towns function solely as central places, there is strong evidence that the urban centres of this area form a four-order hierarchy arranged essentially in accordance with Christaller's *Versorgungsprinzip* or $k = 3$ model. Barrie and Owen Sound together form the highest order in the area (Marshall 1969).[7]

Figure 4.4 includes the trade area boundaries of the major shopping centres (including Toronto) as determined primarily from an analysis of the circulation records of daily newspapers. Each centre's trade area was first defined as the area receiving more newspapers from that centre than from any other centre. The boundaries thus drawn were then spot-checked

7 Figure 4.4 may be compared with Plate 52 of the *Economic Atlas of Ontario* (Dean 1969), which presents a six-order classification of Ontario centres. The atlas plate, with which the writer was closely associated, differs from the present Figure 4.4 in several respects. First, Sarnia, Brantford, and Brockville are each demoted by one order in the revised classification. Secondly, Burlington is treated on Figure 4.4 as an integral part of Hamilton. Thirdly, the five places identified on Figure 4.4 as having 'limited regional influence' were regarded in the atlas simply as third-order centres. These changes arise from the fact that the atlas classification was based almost exclusively on retail sales volumes, whereas the classification on Figure 4.4 reflects a more detailed consideration of intercity patterns of dependence. The population and retail sales data used in the preparation of Figure 4.4 were taken from the *1966 Census of Canada.*

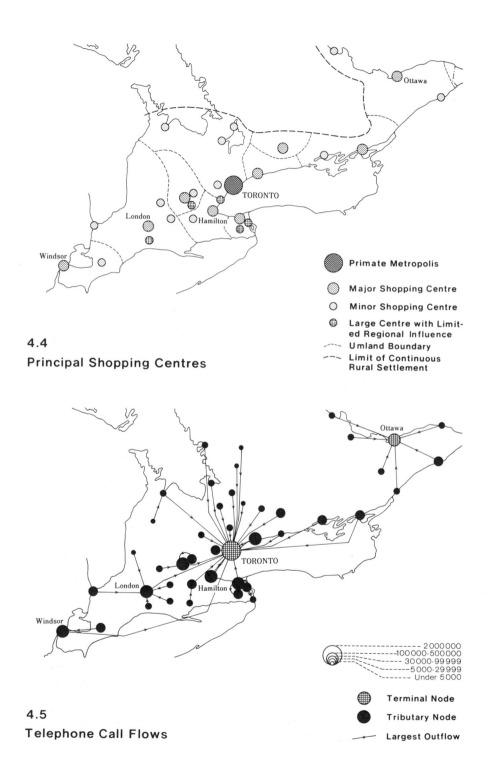

4.4
Principal Shopping Centres

- ⬤ Primate Metropolis
- ◉ Major Shopping Centre
- ○ Minor Shopping Centre
- ⊕ Large Centre with Limited Regional Influence
- ‑‑‑ Umland Boundary
- ‑‑‑ Limit of Continuous Rural Settlement

4.5
Telephone Call Flows

- 2 000 000
- 100 000-500 000
- 30 000-99 999
- 5 000-29 999
- Under 5 000

- ⊕ Terminal Node
- ⬤ Tributary Node
- → Largest Outflow

in the field. Only in the case of Oshawa was newspaper circulation found to be a poor indicator of the area dominated by the city for shopping purposes. This discordance presumably results from the fact that the Toronto papers compete much more effectively with the Oshawa paper in the east than with the papers from the larger and more independent cities of Hamilton, Kitchener, and London in the west. Indeed, the western boundary of Toronto's trade area is accurately delimited by the newspaper data, despite the fact that Toronto is a first-order as well as a second-order centre. To the east, however, the trade area of Oshawa has been enlarged on Figure 4.4 to give a more accurate impression of the city's influence as a shopping centre.[8]

To conclude, some comments will be made on the status of Toronto as Ontario's primate metropolis. In round figures, Toronto contains two million people, while the median populations of the major and minor shopping centres are 125,000 and 32,000 respectively. Toronto is almost exactly five times as large as Hamilton, the province's second-largest city. A general explanation of Toronto's primacy can be formulated in terms of a combination of the 'initial advantage' argument (Pred 1965), and the 'point of entry' concept (Rose 1966). As Gentilcore has noted in chapter 2 above, Toronto was neither the first nor the only point of entry into Southern Ontario for European settlers, but it led in this respect from at least as early as 1835 (Goheen 1970, pp. 44–53). Toronto's harbour, and its selection as the capital of Upper Canada, established the tiny settlement as the point from which a very large proportion of the occupation of Ontario would be controlled. As the nineteenth century progressed, and the economy became increasingly commercialized, it was natural enough that the principal point of entry should emerge as the region's leading marketplace. Since Toronto was now the largest city, it offered the best market and labour supply for manufacturing, and the developing road and rail networks served to crystallize the city's comparative advantage as an industrial centre (Kerr & Spelt 1965, especially chaps. 3 and 5). By the end of the century, Toronto was rapidly approaching a population of a quarter of a million, and its continued primacy within Ontario seemed already firmly assured (see chap. 2).

An idea of Toronto's present position in the business world may be gained from Figure 4.5, which is taken from a recent analysis of telephone calls in Ontario and Quebec (Simmons 1970). The places mapped are

8 Data on newspaper circulation were made available by the Toronto office of the Audit Bureau of Circulations. The general reliability of newspaper circulation data as an indicator of shopping areas is affirmed by Chatelain (1957) and by Smailes (1966, pp. 132–3).

the toll centres through which long distance calls are monitored. Toll centres serve regions of varying size, but in most cases the toll centre itself is the principal generator of telephone traffic. The arrows refer to 'business' calls (that is, total calls minus calls originating at a place of residence), and indicate the destination of the largest outflow from each toll centre on a typical business day in the spring of 1967. In the event that a centre's largest outflow goes to a smaller centre, the arrow is omitted, and the centre of origin is identified as a terminal node in the network.

If the analysis is kept within Ontario's boundaries, Toronto and Ottawa are the only two centres to emerge as terminal nodes. The largest outflow from Ottawa goes in fact to Montreal, and the addition of this link would place eastern Ontario squarely in Montreal's sphere of influence, leaving Toronto as the only terminal node in Southern Ontario. Toronto's largest outflow goes to Hamilton, making these two cities a reciprocal pair at the core of Ontario's industrial heartland. Toronto's overall dominance is clearly evident from the map. It is also interesting to note the strong similarity between Figure 4.5 and Figure 4.4 with respect to the spheres of influence of London, Windsor, Kitchener, St Catharines, and Ottawa.

Summary
This brief examination of city-centred regions, coupled with the materials presented earlier, serves to identify the Golden Horseshoe as the heartland of urban Ontario. Cities experiencing rapid growth (Fig. 4.1), and cities specialized in manufacturing (Fig. 4.3), are concentrated in the Golden Horseshoe, and here also is found the dominant metropolis of Toronto (Figs. 4.4 and 4.5). The contrasts in growth and economic functions between the Golden Horseshoe and the remainder of the province suggest that urbanization in Ontario may validly be conceptualized in 'core-periphery' terms. Such a dichotomy, however, is somewhat coarse, and further research is needed before the finer details of structure, function, and development in the Ontario urban network are satisfactorily understood.

5 Toronto: Focus of Growth and Change

J.W. SIMMONS AND L.S. BOURNE

The story of Toronto parallels the story of Ontario. From its earliest pioneer beginnings, through the period of agricultural settlement and during the gradual emergence of an integrated commercial and industrial economy, Toronto has intimately shared the fortunes of the province. As we approach the latter third of the twentieth century the two become less and less distinguishable. The processes of concentration of population, economic activities, and decision-making units have now drawn over 50 per cent of the provincial economy into the greater Toronto area. Parallel improvements in transportation and communication systems, and in personal mobility, have brought every other part of the province into close contact with this urban core.

The process, however, is not perfectly symbiotic. Complaints are heard that the economic and social benefits of rapid development are greater near the nexus of growth and control. Nor is the continuation of this process inevitable. Public policy, changing lifestyles, or technological innovations could lead to the decentralization of provincial socio-economic distributions into smaller, less specialized regional units with greater independence.

INTRODUCTION

This chapter is concerned with the form and path of this symbiotic relationship and the impact of growth and change on the form and structure of metropolitan Toronto. Our purpose is to convey an image of the strength of regional integration; to identify the components of the rapid growth experienced in recent years, emphasizing those which are unique to Toronto; and to summarize their reflection in a dynamic urban landscape.

Is Toronto's growth experience different? Traditionally, Toronto has been described as a British city on American soil (Goheen 1970). To understand its changing geography is to understand that its institutions have been and still are transplants from the old country, yet its form, appearance, economy and lifestyles are North American (Kerr & Spelt 1965). In postwar years, the image of the metropolis has been further altered by high rates of foreign immigration, the highest on the continent; rapid economic and population growth overall (population has doubled from 1.2 million in 1951 to 2.4 in 1971); revisions in urban governmental forms and in transport priorities; and a scale of redevelopment which suggests to the observer that a complete restructuring of the landscape is in process (Table 5.1).

Contrasts with American cities are evident: the relative absence of intense racial and territorial frictions; a degree of co-ordination in development through a new (1953) metropolitan governing and financial structure (Smallwood 1963; Kaplan 1967); an early postwar start on a subway system; and a continued healthy state of inner-city neighbourhoods. Although there are many parallels between recent changes in Toronto and that of its American and even European counterparts, there is sufficient reason to emphasize specific aspects of the translation of growth processes into different urban forms.

Delimiting the Metropolis
The increasing integration of Toronto and Ontario means that any metropolitan boundary is diffuse and blurred. Many different delimitations of Toronto exist (Fig. 5.1), each with merits and deficits for certain purposes, but none do justice to the complex and evolving relationship between Toronto and the rest of the province, nor to the diversity of change within the metropolitan landscape.

As a basis for the discussions of growth to follow, a full range of concepts of Toronto as a spatial entity has been identified (Table 5.2). Each is at a different scale and each nests within the larger scale. The terminology, of course, is defined with respect to Toronto, so that the relationship with Ray's heartland–hinterland concepts in chapter 3 are not exact; although the 'urban field' in Table 5.2 would approximately coincide with the Toronto portion of the 'heartland.' Four concepts are particularly useful.

1   *The central area* includes the historical city centre and the present central business district and surrounding fringe. Land uses are predominantly commercial or institutional in the core and highly mixed, including residential, in the fringe. At present the core is bounded by the CNR tracks, University Avenue, Davenport Road, and Jarvis Street.

2   *The urbanized area,* as in standard use, is the relatively continuous

Table 5.1 Comparative rates of growth in population and building activity: Toronto and selected United States metropolitan areas*

Metropolitan area	Population (000s)		Population change % 1960–70	Per capita building† permits (1969)	Housing starts per 1,000 persons (1969)	Population change % 1950–60
	1960	1970				
Toronto, Ont.	1,824*	2,368	33.0‡	$247	14.6	50.7‡
Washington, D.C.	2,002	2,875	38.4	108	9.4	36.7
Montreal, Que.	2,109*	2,570	23.2‡	104	8.7	43.3‡
San Francisco, Calif.	2,649	3,068	15.8	122	7.1	24.0
Los Angeles, Calif.						
Long Beach, Calif.	6,038	6,970	15.4	111	6.0	45.5
Chicago, Ill.	6,220	6,894	10.8	99	7.5	20.1
Detroit, Mich.	3,762	4,162	10.6	97	6.1	24.6
Philadelphia, Pa.	4,342	4,774	9.9	63	4.9	18.3
Cleveland, Ohio	1,909	2,044	7.0	89	6.2	24.7
New York, N.Y.	10,695	11,410	6.7	50	3.2	11.9

*1961 figures for Canadian cities.
†Non-residential construction only.
‡Estimated 10-year growth rate.

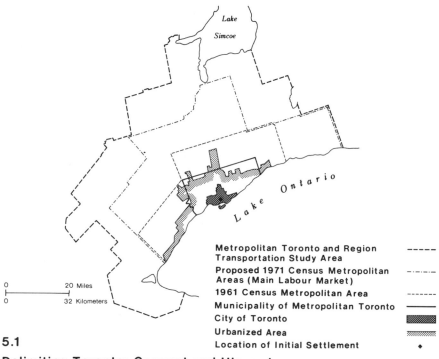

Metropolitan Toronto and Region
Transportation Study Area

Proposed 1971 Census Metropolitan
Areas (Main Labour Market)

1961 Census Metropolitan Area

Municipality of Metropolitan Toronto

City of Toronto

Urbanized Area

Location of Initial Settlement

5.1

Delimiting Toronto: Concept and Hierarchy

Table 5.2   Definitions of scale

Spatial entity	Area (sq. miles)	Population	Scale	Criteria
Province	344,000	7,550,000	—	Political
Hinterland	200,000	6,500,000	4	Service
Urban field	3000	3,500,000	3	Travel
Urbanized area	300	2,350,000	2	Land use
Metro	250	2,000,000	—	Political
Central city	35	670,000	—	Political
Central area	5	115,000	—	Function
Core	1	5,000	1	Density

expanse of land built up for urban purposes. At present its boundary is roughly approximated by the municipality of Metro Toronto (Fig. 5.1).

3 *The urban field* is the area utilized by urban residents for frequent outings. Within it is the urban fringe, arbitrarily defined here as the area within an hour's travel time of the urbanized area. The limits are somewhat larger than the normal daily commuting zone. The urban field bears the brunt of the growing metropolis; it extends as far as Bowmanville on the east, Barrie on the north, and Hamilton on the west.

4 *The hinterland* designates that broad zone which Toronto dominates in providing higher-order services. It encompasses virtually all of the developed areas of the province, with the exceptions of Ottawa and eastern Ontario which are linked to Montreal, the extreme north and northwest which is in the service area of Winnipeg, and possibly the Lake St Clair area surrounding and oriented to Windsor–Detroit. Growth within the hinterland is the subject of most other papers in this volume, and will receive only passing attention here. Yet the form of growth that takes place is articulated by the area's close links with the fringe, urbanized area, and core of the metropolis.

Each of these four scales of measurement defines geographic areas as well as different mixes of growth forces. Subsequent sections of this chapter examine these different mixes and their impact, emphasizing the urbanized area and core, in terms of land use and social change.

Growth Trends: An Overview
The evolution of the symbiotic Toronto–Ontario relationship has been summarized in chapters 2 and 3. The rapid growth of the province in the last one hundred and fifty years has resulted in a continuous centralization of Ontario's economic activity in metropolitan Toronto and its fringes. This concentration is reflected in relative rates of population growth. Toronto's growth has been faster and more stable than that of the province since 1871; its share of total provincial population has risen from 7 to 31 per cent in the last one hundred years. In terms of economic indexes, Toronto's share of the total is even higher. In recent years, Toronto has secured nearly half of all new immigrants and over 50 per cent of new housing and industrial building permits in the province. Over 80 per cent of all cheques are cleared through Toronto brokerages.

Waves of in-migrants, combined with high rates of natural increase, have been long-term common denominators of growth. The arrival of immigrants from other parts of Canada and abroad has been a characteristic of the city for some time, although the impact is particularly striking at present because the immigrants differ sharply in character from the traditional Toronto stock (Fig. 5.2).

The most obvious effect of such continued growth is that social and physical change are very much the way of life. Newcomers modify the cultural mix, adding new preferences and new ways of doing things. New buildings rise, communities turn over rapidly in ethnic composition, and new institutions are created in response to new social demands. Consequently, there has been and still is a continuous evolution of the city's lifestyle and self-image – the way its citizens and outsiders see it. The frontiertown, the Tory bastion, the dour Protestant city of churches, and the cosmopolitan metropolis are all roles which Toronto has cherished at one time or another in its history (Arthur 1965). But against these popular images of vitality, opportunities, and social mobility, the innovations and stimuli brought about by growth, and the interaction of different cultures must be weighed the realities of massive and crude landscape alterations, quick-and-dirty construction, the stress of culture shock, the uprooting of neighbourhoods, and the rapid erosion of traditions and values (Lorimer 1970).

With continued immigration, demographic, ethnic, and religious differences have steadily become more complex (Fig. 5.2). The city has recently become more Catholic and more Latin. Various ethnic groups are now of sufficient size to create cities within the city. An estimated 400,000 persons of Italian origin inhabit a complete sector of the city from the core to the northwest suburbs; a parallel sector contains 100,000 Jews, and other areas house 50,000 Greeks and nearly as many Portuguese. Each of these subcities, and there are many others, contains its own commercial core, theatres, institutions, and social structure. Tastes vary widely from one such ethnic city to another; and as a result the range of behaviour over the entire social spectrum has expanded in concert.

Other demographic patterns are similarly sensitive to immediate past growth patterns. National trends in family size and longevity are reflected in the postwar shifts in Metro Toronto's age structure. In-migrants are predominantly young people. Toronto has a strikingly high proportion of young singles and newlyweds – often employed in professional and clerical occupations and demanding of the conveniences provided in modern rental accommodation, particularly high-rise apartments. They are geographically as well as socially mobile, discriminating in their needs for services and fluid in their demands. The economy and landscape mirror their presence – the increasing diversity of occupations, of night life and movie houses, of massive and impersonal apartment developments with built-in services, and the growing vitality of the core.

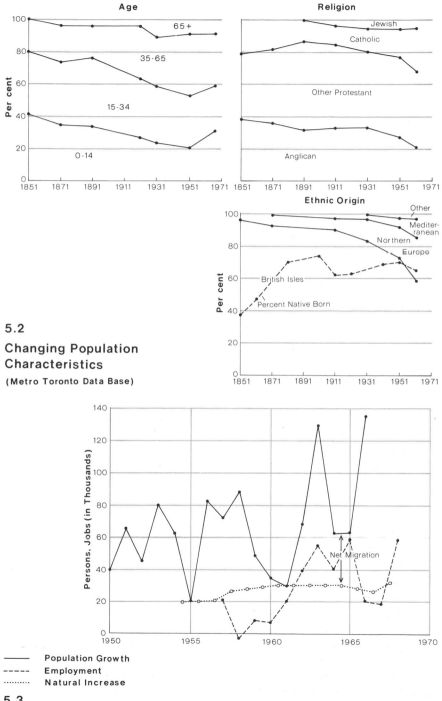

Age

Religion

65 +

35-65

15-34

0-14

Jewish

Catholic

Other Protestant

Anglican

Ethnic Origin

Other

Mediter-
ranean

Northern
Europe

British Isles

Percent Native Born

5.2

Changing Population Characteristics

(Metro Toronto Data Base)

Net Migration

——— Population Growth

----- Employment

·········· Natural Increase

5.3

Recent Yearly Fluctuations in Growth

Although the long-term growth profiles for Toronto described above are quite regular, actual growth rates fluctuate widely from year to year (Fig. 5.3). Metropolitan growth reflects broad cyclical changes in international, national, and provincial economic conditions, but it actually takes place by means of innumerable individual decisions. These decisions are based on people's knowledge of the present and perception of the future. Developers build, entrepreneurs invest, families have more children, and finally government reacts with the provision of new services, but any of these decisions can be accelerated or delayed over time – producing wide variations in growth rates from year to year.

Of all the components of growth, natural increase is the most stable. Although currently adding about 30,000 persons to the Census Metropolitan Area each year, recently the rate of increase as elsewhere has slackened. Natural increase per 1000 has dropped from 17.7 in 1960 to 9.7 in 1969 in the province. To this is added (or subtracted) net migration, which in Toronto has ranged from 0 to nearly 100,000 persons per year during the last decade, averaging around 30,000 but displaying a distinct cyclical behaviour.

These migration cycles are integrated in complex ways with the family of development cycles characteristic of a growing metropolis. Net migration is a response to the creation of new jobs, generally lagging behind the latter by about a year. Approximately 12 to 18 months after population growth occurs through in-migration, the housing stock is altered as new units are begun. Somewhat later, the public sector responds with additional social services. The creation of jobs occurs hand in hand with investments in buildings in the industrial, commercial, or governmental sectors. The Toronto landscape, to which we now turn, is one distinctive product of this growth history.

THE IMPACT OF GROWTH AND CHANGE

The concentration of Ontario's growth and investment in Toronto has required a continuing and rapid adjustment in the spatial structure at each level of scale of the Toronto-oriented region: the central core grows and becomes more specialized; the urban area expands outward and modifies its internal structure, while activities in the urban fringe and hinterland become more closely linked to the urbanized area. New factories and homes on the margins of the built-up area are accompanied by the redevelopment and replacement of older activities in the centre, and by adjustments in the transportation, utility, and communication systems that are required to serve a larger, denser population (Fig. 5.4).

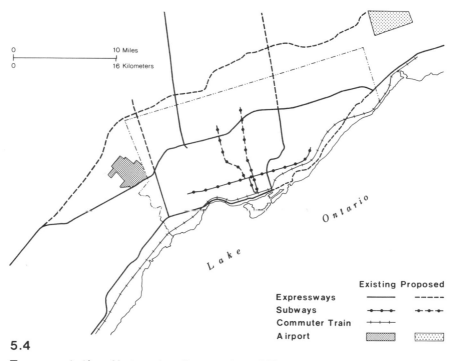

	Existing	Proposed
Expressways		
Subways		
Commuter Train		
Airport		

5.4

Transportation Networks: Present and Future

As in most cities, the aggregate expansion of the Toronto urbanized area over time reveals a regular octopodan spatial pattern (Fig. 5.5). The detailed form of change in the urban margin depends, of course, on building and transportation cycles, and social preferences prevalent at that period in history. Such details are beyond the scope of the present volume. It is apparent, however, that the scale of dispersion has increased substantially in postwar years. New growth has leap-frogged outward to envelop small towns and cities in the fringe. As the urban margin expands in pursuit, outliers of earlier growth around villages and towns are subsequently swallowed up, adding complexity to the resulting urbanized landscape. The composite pattern is further modified by urban geomorphology (in Toronto ravine lands exert a prominent influence) and by the historical inheritance of earlier decisions on transportation and other service investments.

Changes in Land Use

Figure 5.5 does not, however, reveal the critical patterns and rates of intra-urban land use change that are the response to new growth. The

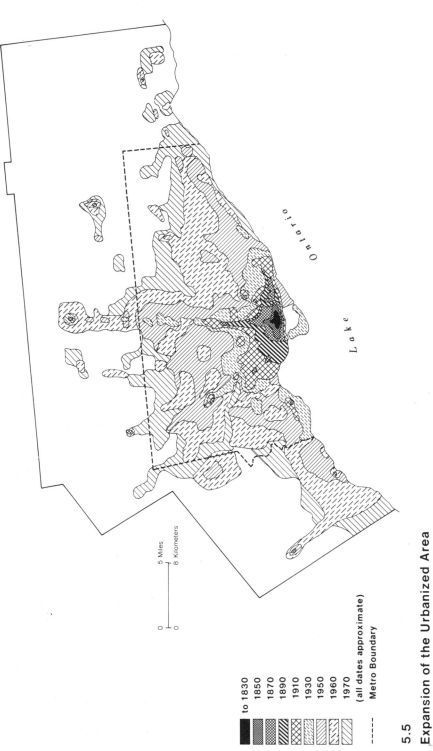

Lake Ontario

to 1830
1850
1870
1890
1910
1930
1950
1960
1970
(all dates approximate)

Metro Boundary

0 ⊢ 5 Miles
0 ⊢ 8 Kilometers

5.5

Expansion of the Urbanized Area

most obvious change is the transfer of land from agricultural to urban use. The rate at which this consumption occurs depends on population growth totals and on marginal rates of per capita land requirements. Table 5.3 summarizes recent changes in both aggregate and per capita use of land in Metro Toronto. With a particularly high growth rate, Toronto has shared the experience of most other cities during the postwar period in recording continually higher rates of land consumption.

The internal rearrangement of land uses takes place through the process of succession. Low-density uses are replaced by higher-density uses; single-family homes are torn down to make way for apartment buildings; institutions and commercial activities expand at the expense of neighbouring uses; parking lots appear where old shops and houses once stood. Despite the visual chaos, there is regularity to the succession process. In Toronto, a rapidly changing inner-city produces a complex array of transitions from one use to another. This complexity is captured in Table 5.4 which records the proportions of land use shifts over ten years for the city of Toronto. Note the major sequences of change: the shift to higher-density uses, the heavy burden borne by residential areas, the substantial movement into parking and even vacant land, the relative absence of succession between certain mutually exclusive uses such as industrial and residential. Yet, despite these significant shifts, most areas remain in the same category of use (main diagonal in Table 5.4) although the density of that use may increase through redevelopment. Thus existing distributions of uses remain basically stable.

Aggregate measures, however, describe only part of the story of rapid change. The spatial patterns of net land use change, reflected in the above numerical transitions, are equally complex. When net changes in both the suburbs and in built-up areas are combined and factor analysed, four rather than two distinct spatial patterns are identifiable (Fig. 5.6). These are: growth of the urbanized area margin through the conversion of vacant or agricultural land to low density urban uses; expansion of the core area through the redevelopment of older commercial and residential uses by new offices and, on a more temporary basis, by parking lots; infrastructure growth, consisting of additions to and revisions of the networks of urban expressways and public utilities; and major nucleations of activities including government office complexes; large new institutions such as hospitals, a suburban university and community colleges; suburban hotel clusters around the airport and along highway arterials; and extensive high-rise apartment projects next to the commercial core, or adjacent to farmers' fields on the urban periphery.

Each of the above processes provides an independent and additive

Table 5.3   Land use and land utilization, Metro Toronto,* 1963–8

| Population growth 298,000 | Land use | | | | | | Utilization | | | |
| | 1963 | | 1968 | | Change | | Acres/1000 population | | | Change/1000 pop. increase acres |
	Acres	%	Acres	%	Acres	%	1963	1968	Change	
Commercial	4888	3.2	5610	3.6	722	14.8	3.0	2.9	− 0.1	2.4
Institutional	5873	3.8	7795	5.1	1922	32.7	3.6	4.1	0.5	6.5
Open space	20,924	13.6	21,367	13.9	443	2.1	12.9	11.3	− 1.6	1.5
Industrial	10,342	6.7	12,516	8.1	2175	21.0	6.4	6.6	0.2	7.3
Residential	53,936	35.1	60,223	39.2	6287	11.7	33.3	32.0	− 1.3	21.1
Transport and Utilities	11,257	7.3	12,277	8.0	1010	8.9	6.9	6.5	− 0.4	3.7
Undeveloped	46,420	30.2	34,012	22.1	−12,400	−26.7	28.7	18.1	−10.6	—
Totals	153,640	100.00	153,802†	100.0	—	—	—	—	—	—

*Municipality of Metropolitan Toronto.
†162 acres were added by filling along the lakeshore.

Table 5.4 Land use succession: transition probabilities for city of Toronto, 1952–62 (uses ranked by density)

Existing use (ranked by intensity)	1	2	3	4	5	6	7	8	9	10	Total no. of parcels redeveloped	No. of parcels redeveloped in same use
1 Office commercial	0.93	0.01	0.01	0.01					0.03	0.01	123	49
2 High-density residential*	0.01	0.95							0.01	0.03	32	10
3 General commercial	0.02		0.92					0.01	0.04	0.01	573	156
4 Warehousing			0.04	0.84	0.06			0.01	0.03	0.01	313	63
5 Industry			0.01	0.05	0.90			0.01	0.02	0.01	216	68
6 Low-density residential		0.03				0.94			0.02	0.01	3114	132
7 Transportation	0.01	0.01					0.94	0.01	0.03		22	1
8 Auto commercial	0.09		0.05					0.80	0.03	0.03	172	115
9 Parking	0.09	0.03	0.04	0.02		0.04	0.01	0.06	0.76	0.05	253	81
10 Vacant	0.02	0.05	0.02	0.15	0.09	0.17		0.03	0.11	0.34	2346	—
No. of new parcels	615	713	503	383	292	1424	10	512	1821	502	7164† 6775†	675

NOTES Rows sum to 1.00 except for minor rounding errors. Blank cells have values of less than 0.005. Observations are individual properties or parcels.
*Housing with more than 6 units.
†Number of new properties does not equal the number of parcels redeveloped as the former excludes construction in progress.

component in the pattern of land use change. The first two patterns appear as rough rings of rapid change around the central core, with suburban growth taking place about 10 miles (and 80 years) ahead of redevelopment of the core. In each of the rings as much as 30 per cent of the land area within a given census tract has been altered in the short five-year time span. In other words, the unusual vitality of the core, and of the inner-city generally, has meant that some areas of the built-up landscape of central Toronto are being renewed at a rate equal to the rate of expansion of the suburban fringe.

Evolution of Social Areas

Analogous patterns to those of land use change are evident in the evolution of social areas within the city. The rapid growth of Toronto has added new populations with quite diverse socio-economic characteristics. The high proportions of young in-migrants and of immigrants from southern Europe, in particular, have substantially and permanently modified the traditional social geography of the city.

Despite social diversity and rapid change, the underlying geometry of social areas in Toronto is similar to that of most North American cities. For example, in his cross-sectional factorial ecology of Metropolitan Toronto based on the 1961 census, Murdie (1969) found that communities could be arrayed on the basis of six independent and additive factors: (1) socio-economic status; (2) family status (urbanization); (3) Italian ethnic status; (4) Jewish ethnic status; (5) areas of recent growth; and (6) household characteristics and service employment. Most of these patterns are common to other cities: the first factor displays concentric zonation, the second a sectoral pattern, and the others are smaller scale nucleations often described by sociologists as reflecting social segregation. The specificity of Toronto's social ecology is represented in two ways: the separation of what is usually a single composite ethnic dimension into two factors, Italians and Jews; and the separation of some residential areas solely by their recent and rapid growth.

More revealing of Toronto's changing character is the analysis of social change itself. Murdie identified four major patterns of social change (Fig. 5.7), peripheral growth, increasing ethnic segregation in housing, 'urbanization' and the zone of residential stability just inside the zone of suburban growth. 'Growth' refers primarily to population change, but is also associated with the increased income levels of families living in new housing. Sociologists use the term 'urbanization' to describe the family-cycle dimension, which varies from a family-oriented suburban lifestyle to the apartment dwelling young people in the city centre. The ethnic segrega-

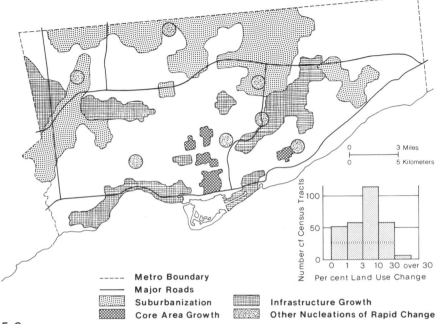

0 — 3 Miles
0 — 5 Kilometers

Number of Census Tracts

100

50

0

0 1 3 10 30 over 30
Per cent Land Use Change

- - - - - Metro Boundary
———— Major Roads
Suburbanization
Core Area Growth

Infrastructure Growth
Other Nucleations of Rapid Change

5.6

Components of Land Use Change: 1963-68

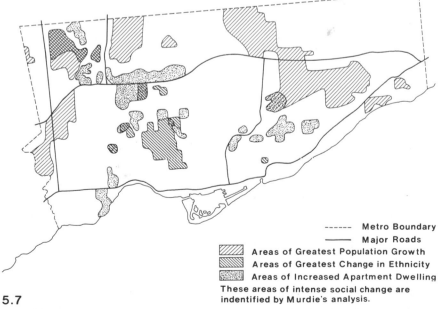

- - - - - Metro Boundary
———— Major Roads
Areas of Greatest Population Growth
Areas of Greatest Change in Ethnicity
Areas of Increased Apartment Dwelling

These areas of intense social change are
indentified by Murdie's analysis.

5.7

Patterns of Social Change: 1951-61

tion scale differentiates areas that are rapidly becoming more Italian and more Catholic.

Growth, Mobility, and Interaction

Most of the patterns of social and structural change noted above derive from the rapid growth of Toronto. Growth generates increased demand for all kinds of space, and the simultaneous increases in the size of many different clusters of activities and social groups lead to adjustments in their location and internal structure and to modifications of the networks of interaction which link them.

These patterns of aggregate change summarize the locational adjustments of individual activities – the movements of stores, expansion of institutions, and creation of new firms. Social change, for instance, can be regarded as the summation of the movements of individual households. About 20 per cent of the population in Toronto moves each year, but only one-third of the movers leave the metropolitan area (Simmons 1968). The spatial patterns of intra-urban migration reflect the impact of growth on a massive human circulation system in which households alter their lifestyle, and their place of residence, as they move through the life cycle.

The cycle is further subdivided by class and ethnic characteristics, and by patterns of social contact. Aside from the very poor, who tend to move about in the older housing around the city core, most families can be allocated in a rough fashion to one of the six major social sectors of Toronto (as previously identified) according to social class, ethnic background and past residential history. Life-cycle adjustments for each family then take place primarily within its own sector of the city (Fig. 5.8).

The urbanized area is a heavy net exporter of households to practically all locations in the urban fringe. Migration between parts of the urbanized area and the fringe and among various centres in the fringe are basically sectoral, as they are within the city. Restructuring of the fringe due to the influence of Toronto may therefore result in a continuation of the socioeconomic sectors so clearly demarcated within Metro itself into the north Yonge Street, Ajax–Oshawa, Mississauga–Oakville, and Brampton–Bramalea corridors.

The intimate relationships between growth patterns – the result of permanent moves of households, activities, and investments, and the day-to-day linkages of persons, information, and goods can be demonstrated in other ways. The relocations of activities reflect past and potential adjustments in functional linkages, and the analysis of the latter adds a further dimension to our knowledge of processes of change in Toronto (Fig. 5.9).

One prominent integrating feature of the Toronto landscape has been

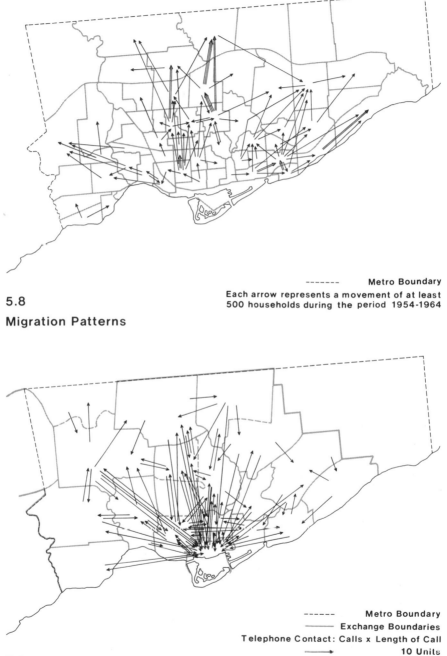

5.8

Each arrow represents a movement of at least
500 households during the period 1954-1964

Migration Patterns

------- Metro Boundary
_____ Exchange Boundaries
Telephone Contact: Calls x Length of Call
———▶ 10 Units
———▶ 3 Units

5.9

Patterns of Interaction, 1967

the continued strength of the central core. In virtually all patterns of interaction and contact – such as phone calls, commercial shopping trips, journeys to work, and so forth – the downtown area and its immediate extensions remain the dominant focus. The vitality of the core indicates a continuing desire for close personal contact, as well as for access to the large pool of clerical help in Toronto's densely populated inner-city neighbourhoods. The clusters of land-use changes identified earlier further emphasize the agglomerative linkages among integrated activities not only in the core but throughout the urban area.

Together with increasing income, leisure time, and mobility, social interaction in general and travel in particular have increased substantially. The definition of the urban field at the beginning of this chapter is suggestive of the integration of the Toronto region through wide-ranging patterns of recreational travel behaviour. By automobile, boat, and snowmobile, the urban fringe is now fully integrated as Metro's 'backyard' and neighbourhood park.

FOCI AND ISSUES OF CHANGE

The rapid growth of a city, and its emergence to metropolitan status, tend to focus interest on certain critical areas of change. One impression created by the preceding discussion is that urban growth, and the cultural and technological evolutions which this growth mirrors, are highly selective in their spatial impact. This section identifies examples of the resulting nodes of intense change within the context of recent changes in Metro's transportation system. Transportation innovations are the starting point because of their obvious importance in shaping patterns of accessibility, which ultimately affect interaction and thus patterns of social and functional change.

Transportation Network

The transportation network simultaneously responds to and generates urban growth. New facilities are built in response to demand – in turn modifying demand by limiting alternatives and concentrating interaction within specific channels. The pattern of major transportation links built and proposed within the province (see Fig. 5.4) gives an excellent picture of past and future growth trends.

Two recent changes in the transportation system have had a substantial impact on the form of Toronto: new regional expressways and urban subways. The expressway network of the province remains heavily focused on Toronto, in fact all such routes touch the metropolitan area (MacKinnon

& Hodgson 1969). Historically the road network has always been oriented to Metro, but the selective and agglomerative impact of recent changes in the system has been of a different order of magnitude. Highway 401 has become Ontario's 'main' street, with Toronto as its peak land value intersection. Most new provincial growth occurs within a few miles of this regional artery. Other significant highway investments have, of course, been made, but many such as those in the Muskoka and Kawartha Lakes recreation areas north of Toronto have served primarily to entrench the integration of the urban fringe with the city and to expand the recreational space of the Toronto population.

Future improvements in transportation will maintain if not enhance this concentration. New provincial expressways, as well as proposed high-speed inter-city rail service and suburban commuter lines will likewise radiate from the metropolis. The proposed new international airport well beyond the metropolitan boundaries also carries with it an enormous multiplier effect in the growth of the fringe.

Areas of Intense Change
Clearly, the evolution of a transportation network inevitably polarizes growth. In Toronto, one effect of recent transportation innovations has been the creation of diverse growth foci. These nodes, in turn, begin to influence the nature and form of subsequent development: of land use patterns, densities, social change, and of other additions to the transportation network. Although any profile of the intensity of change will vary from time to time, it will isolate areas of almost complete turnovers of populations and activities from those of relative, if only temporary, stability.

Within the Toronto region the transportation system has created key growth foci in the urbanized area and around outlying settlement nodes. Highways 400, 27, and the Queen Elizabeth Way, in addition to highway 401 have attracted clusters of high traffic generating activities – large sprawling industries, institutions, subdivisions, and shopping centres. For example, all three of Toronto's new enclosed shopping malls – Yorkdale, Fairview, and Sherway – are adjacent to this network.

Within the Toronto urbanized area, major growth foci also show a close link to the geometry of the transportation network, particularly the subway. The largest node, the downtown core, remains volatile and dominant. Although the area represents only 0.3 per cent of the land area of Metro, it contains 16.5 per cent of all jobs, including 48 per cent of all jobs in finance and nearly 61 per cent of office space. In virtually all measures it is losing relative ground (Table 5.5); yet, unless there is a drastic revision

Table 5.5   Growth foci: growth and replication of the core

District	Employment		Office space (% of Metro Toronto totals)		Apartments	
	(1956)	1964	1953	1969	1951	1966
Downtown core[1]	19.3	16.5	71.7	49.3	—	—
Uptown (Bloor–Yonge)[2]	3.7	5.1	7.8	11.2	4.4	2.9
St Clair–Yonge[3]	1.0	1.1	2.0	6.1	3.6	2.0
Davisville–Yonge } [4] Eglinton–Yonge	1.4	2.7	0.7	6.7	5.3	4.3
Total %	25.4	25.9	82.2	73.3	13.3	9.1
Metro totals	630.2 (000s of employees)	711.8	10.2 (millions sq. ft.)	27.5	60.3 (000s units)	195.2

1 Defined as the area south of College Street, between Simcoe and Jarvis, and north of the CN railroad tracks (Census tracts 73, 74, 75, and 76).
2 Census tracts 70, 71, 72 and 95.
3 Census tracts 66, 67, and 91.
4 Census tracts 84, 85, 86, and 87 (and 303 in 1966).

in transportation priorities, it will likely continue to increase in absolute strength as well as size. Downtown employment is expected to rise, after a period of stability in the early 1960s, particularly if proposals for massive redevelopment of the waterfront (Metro Centre, Harbour City, and Harbour Square) are realized in whole or in part. Within the core, redevelopment has been rapid – the usual combination of office, institutional, and parking uses – adding over 6.0 million square feet of office space in the last decade. In terms of land area, parking has shown the greatest percentage growth. Warehousing and commercial uses, on the other hand, have declined by 50 per cent since 1956. New functional clusters have begun to emerge, both in the core and in its fringe, often linked by pedestrian malls to each other and to subway stations.

Outside of the core, the striking landscape image in Toronto is the proliferation of growth nodes. High-density concentrations of office and apartment developments have appeared at subway nodes, such as at Bloor, St Clair, and Davisville–Eglinton, and in areas near expressway interchanges such as Don Mills–Eglinton. The independent effects of the subway on restructuring the urban area are impossible to isolate, but the reallocation and concentration of development has been substantial. Nearly 50 per cent of all office construction in Metro in the last five years has taken place outside the core, mostly in three districts listed above along the Yonge subway line (Table 5.5). Similar concentrations in high-rise apartment construction, reflecting both the attraction of accessibility as well as policy directives, have sprung up in the same areas as well as along the newer east–west subway line.

Clearly a new landscape is in the making. Encouraged by construction of the subway, and driven by the rapid growth of the city, a variety of high-density foci have emerged within both the inner city and the suburbs. Despite the continued vitality of downtown, a polynucleated urban form is emerging, or has already emerged, with each node developing its own unique and complementary mix of functions, and to a lesser extent of residents. The core has replicated itself.

IMPLICATIONS

The theme of this chapter has been the increasing concentration of Ontario's population and economic activity in Toronto, and the dramatic restructuring of the urban area that has resulted.

Table 5.6 is an attempt to knit together common threads from the preceding discussion. For three levels of generalization the table summarizes the implications of urban growth in terms of the form of development, and changes in land use, social patterning, and transportation facilities. The view we have taken is from the city outward: what changes in the fringe and hinterland derive specifically from or are a response to growth of the metropolitan core?

Among Torontonians there is a growing awareness of the intimate relation between the urbanized area and its surroundings. This may lead to government modification of fringe area growth patterns in an attempt to place a limit on sprawl and to achieve a more balanced mixture of open space and urban development than has been the case in the past, and then it may not. Obviously by controlling most of the new infrastructure support, including education, transportation, water and sewage facilities, such a balance might in theory be achieved. The recent Toronto-Centred Region plan proposing controlled decentralization is one view of what this form might be (Ontario Department of Treasury and Economics 1970). Within the city the public is becoming increasingly aware of the aesthetics of the built environment, criticizing pollution, the scale and disruptive effects of new developments, bureaucratic inflexibility, and the monotony of the automobile-oriented landscape (Lorimer 1970). Such attitudinal changes may be implemented through the market-place by means of consumer preferences, or through pressures on public policy. The total impact

Table 5.6 Spatial implications of change: a summary of the Toronto experience

Component	I (Urbanized area)	II (Urban fringe)	III (Hinterland)
		Scale of measurement	
Form of development	(a) growth at boundary	(a) spill-over pressures from expansion of area I	(a) growth at major nodes only and proportional to size and distance from Toronto
	(b) redevelopment within	(b) growth at all nodes, many are gradually enveloped as urbanized area expands	(b) random growth at certain resource sites – unstable
Land use	Four major patterns: (a) core redevelopment and replication	(a) structuring of ownership and use with respect to Metro market	(a) structuring of economic activities with respect to Metropolitan core
	(b) fringe expansion	(b) transfer of agricultural land to residential and recreational use	(b) replacement of production priorities by consumption (recreation) priorities
	(c) revisions of transport and utilities networks	(c) decentralization of certain urban activities	
	(d) growth of major nucleations – institutions, commercial, apartments	(d) political reorganization – regional authorities	

Table 5.6 Continued

Component		Scale of measurement	
	I (Urbanized area)	II (Urban fringe)	III (Hinterland)
Social areas	(a) continued population growth, greatest at periphery	(a) rapid growth originating from urbanized area	(a) population growth – with associated demographic, ethnic and social class changes – concentrated at a few major nodes
	(b) ethnic shifts dominate internal changes as social areas settle	(b) young urban families modify older rural population structure	(b) continued net out-migration from rest of region
	(c) increased life cycle and ethnic segregation through aging and urbanization processes	(c) sectoral social class patterns of urbanized area superimposed	(c) pockets of rural poverty develop
Transport systems	(a) expressway system surrounds core	(a) expansion of commuter services to downtown core	(a) Toronto-centred expressway network evolves
	(b) subways focus on core	(b) expressways identify growth nodes	(b) major highways to consumption (resort) areas
	(c) peripheral highway networks emerge	(c) extensive single-purpose facilities: major airports and railroad freight yards	(c) reduced alternatives in transport facilities within local areas

of this awareness, and how long it will last, is difficult to predict, but it may lead to reduced rates of overall growth, more selectivity in the location of redevelopment, a drastic revision of priorities between transportation facilities and social service investments, and greater political sensitivity to social needs. If so, the future could witness slower and more purposeful growth in which the difficult trade-offs between expansion and stability are more fully recognized.

In summary, Toronto's recent growth obviously presents many dilemmas, some of which are unique. One is the attractiveness of the central area for redevelopment. Is it possible to accommodate growth as well as to maintain some confidence and stability in inner-city residential areas? A second contrasts the desire to limit continuous sprawl of the urbanized margin with the costs of forcing new growth far beyond the metropolitan boundaries. Another asks what are the mixed consequences of continued rapid population growth, of a high rate of diverse foreign immigration, and of rapid social turnover on the city's economy, culture, and ways of life? Finally, what are the future costs and implications of attempting politically to 'design for tomorrow' a different urban structure than is presently evolving? Beyond our studies of process then must lie investigations of attitudes and preferences.

6 The Political-Territorial Structure

C.F.J. WHEBELL

The first chapter of this monograph outlined the major physical constraints that man has had to deal with in achieving the present-day spatial organization of the province. Next the pattern and timing of the spread of Europeans onto the land with the accompanying rise of urbanism was described (chapter 2) and the resulting contrasts between the urban core or heartland of Ontario and the hinterland or remainder of the province was analysed (chapter 3). Then the current tendencies amongst the urban places of Ontario (chapter 4) and in the central metropolis in particular (chapter 5), were examined. Accompanying all of these changes were changes in the mode and functions of political–territorial organizations. But the changes were not made easily or smoothly; there were and are sources of political stress in the periodic and frequent attempts by both official government and non-governmental interest groups to adjust the political structure to the current pattern of population, economic activity, and political power.

In the earliest stages, when most of the energies of the population were devoted to extending land settlement in the traditional mode of the frontier, the political emphasis was on decentralization from those very few centres of administration and decision-making characteristics of the quasi-autocratic governments of the eighteenth century to the many centres expressive not only of the philosophy of 'grass-roots' democracy but of the realities of the frontier. In this context, communities of interest were normally quite small both in numbers and territorial extent, reflecting the difficulties of travel and communication in an underdeveloped land. More recently, as both communities of interest and the effectiveness of transport have grown, and as the demands of an increasingly urbanized and technologically demanding society on its local-government institutions have outstripped the capability of these institutions to meet these demands, the emphasis has shifted to a process of recentralization.

In the beginning of official settlement in what is now Ontario in 1784, the pioneers came from parts of North America, lately seceded from the

First British Empire, in which a strong tradition of local democratic institutions had developed in the preceding century and a half. For the first few years, the settlers were under quasi-military organization, and administration was effected locally by the officers of each township. But for purposes of judicial procedure, including the prosecution of civil suits, the settlers were at a great distance (some as much as 500 miles) from the westernmost place (Montreal) at which such issues could be resolved. As well, most of these settlers resented the fact that they were expected to carry on the process of clearing and developing the land – an extremely arduous one in the circumstances – with no real 'ownership' of the land or of their labour investment, at least in the sense to which they had become accustomed, since they were, in effect, under French law established by the Quebec Act of 1774.

The solution to these two problems was found in decentralization. The first step in this direction was undertaken in 1788, when four districts, created in what is now Ontario, allowed some judicial and administrative problems to be decided locally, within the policy framework of regulations and instructions issued from the Governor at Quebec. Each of the districts was centred on one of four original settlement areas (chapter 2), which functioned in this respect as a political core area. The administration of each district was in the hands of magistrates, selected by the Governor from among the men of substance and political reliability (often ex-officers) in each settlement area. The boundaries of the districts were simply defined, in accordance with both the simple human geography of the period and in the prevailing ignorance of the natural landscape, by straight lines northwards from recognizable places on the shore of the lake and river system that was at the time the sole means of communication among the widely scattered core areas. From 1788 to 1791, the British colony of Canada (or Quebec) functioned as a two-tier system, but the lower tier had no decision-making powers at all, being only a set of local agents for the implementation of decisions and policies decided by the Governor (and ultimately by the British Privy Council in Whitehall).

A second move in decentralization, of a different order, took place in 1791, by which the desires of the English-speaking settlers of the four upper districts as to the kind of legal system – and especially the land-tenure system – under which they would live were acceded to. This procedure, implemented by the Constitutional Act of 1791, can be regarded as a successful separatist movement, as the Province of Upper Canada thus created was the forerunner of Ontario, in both the cultural–legal sense and in its territorial basis. Although the King's representative in Upper Canada, a Lieutenant-Governor, was formally subordinate to the

Governor at Quebec, in practice he had a great deal of discretionary power. He was the executive head of a system with real decision-making authority, as Upper Canada was set up with a Provincial Assembly that was modelled as closely as possible on the system prevailing in Britain at the time (Craig 1963). And so the creation of Upper Canada and the provision for local representative bodies was indeed an act of separatism – from the earlier 'mother' colony of Quebec – and also one of decentralization within the British Empire, since a measure of self-government was awarded to the two Canadian provinces so formed in 1791. To be sure, the powers of these assemblies were not extensive, and at all times the King's representatives kept a close eye and a firm hand on the procedures of the Assemblies, but the process of decentralization was well in train.

The districts as formerly set up in 1788 continued to function in approximately the same way as they had before 1791, although it was now the provincial government that directed the deliberations of the magistrates. To carry out one of the provisions of the Constitutional Act, however, another set of units was required. The first Lieutenant-Governor of Upper Canada, John Graves Simcoe, conceived of a set of 'counties' which were to be the basis for the constituencies of the Assembly. In addition, they were to act as a somewhat decentralized division of the districts for the registration of land titles, an extremely important function of government now that freehold titles were possible, and as the basis of organization for the militia or home-defence forces raised from among the citizenry. These counties, however, were never very significant functionally.

While the higher-level changes were being discussed, debated, and implemented, at the level of the local community another institution was making its appearance. The settlers, especially those from New England, were accustomed to meeting together once a year to deal with problems affecting their own locality, to resolve internal stresses and to agree on regulations for the smoother functioning of their own community. The territorial unit that reflected this sort of community in New England was called the 'town,' and the annual gathering to sort out problems and resolve them the 'Town Meeting.'

At the very beginning of settlement in the 'upper districts' there was a need for locating and describing for legal purposes the parcels of land granted to the incoming settlers, and a system of land survey was adopted that involved superimposing a grid of roads and property units ('lots') on the landscape. The term used for this unit for land accounting is 'township,' but it must be pointed out that, unlike the meaning of the word elsewhere, there is no connotation whatever of nucleated settlement. Indeed, there had been in the 1780s a deliberate attempt to prevent these units of land survey from acquiring the character of a local community, in the form of an express order forbidding the use of names to identify townships, which were numbered at first. But the desire of the local inhabitants for some sort of identity asserted itself in the appearance, by the mid-1780s of names for many of the townships in the eastern part of the province; the names were astutely chosen to forestall objections from the authorities on grounds of disloyalty, and included the names of the thirteen children of King George III.

Despite the intentions of the authorities, then, to prevent the rise of local loyalties tied to names of territorial units, it was probably inevitable that the 'townships' would be seen as not merely a survey unit, but the territorial unit for local community expression, and in 1790, the first 'Town Meeting' known was held in the township of Adolphustown (Shortt 1902). Why it took six years from the first settlement to reach this stage it is difficult to ascertain, but apart from the fact that such meetings were proscribed, and the action of the residents of Adolphustown was strictly illegal at the time, it is probable that only when the development of the settlement had reached a certain stage of growth with a sufficiently large and dense population, and problems such as animals straying into gardens had become serious was there a real need for self-regulation at this level.

The widespread desire for this kind of local institution was recognized by the Executive of Upper Canada in 1792, when permission was granted for Town Meetings to be held for very specific – and trivial – purposes, always under the control of the magistrates of the districts, in the same way that they were in turn responsible to the provincial Executive and so on upwards in the hierarchical administration of the Empire (Shortt 1902). And so, under Simcoe's administration, Upper Canada functioned with three levels of government below that of the province itself: the township, the county, and the district. This arrangement was not, properly speaking, a three-tier system, as each level was subject to control from the provincial Executive, and decentralization of authority was not in a truly stepwise fashion.

Further, the units that comprised the three levels were independently defined and laid out in the landscape. It has been noted already that the districts were defined by astronomical boundaries. The counties were defined by boundaries that also were straight lines in part, but most were on azimuths other than north. Finally, the townships were laid out on a variety of bearings, since most of these were based on an approximation of the bank of the St Lawrence, the shores of Lake Ontario, and a few interior baselines that had been surveyed empirically (Fig. 6.1) with reference to known features such as river courses and trails. The resulting confusion of the three sets of territorial units is indicated by the plaintive tone of the despatch of President Peter Russell to the absent Governor in 1797:'The consequence is that some Counties being in two Districts and several Townships in two counties and Separate Districts, the Inhabitants of such divided Counties and Townships cannot decide to which jurisdiction they belong, and of course are neither assessed nor enrolled to the great hindrance of a due organization of the Province ...' (Cruickshank 1923). The correction of this situation, which had been aggravated by the development of new settlement nodes in the spaces between the earlier set, was accomplished by a new arrangement which came into effect in 1800. This Act created an additional four districts by partitioning the original ones, and introduced the 'nesting' principle by making all higher-order units conterminous with the boundaries of townships, which have ever since been treated as territorial modules in the development of the political system. Although not much advance took place at this time or during the next forty years in the functions of the townships as local-governmental institutions, at least they remained in being as the lowest-order political units in the province.

Subsequent to 1800, the settlement system continued to develop, largely through the formation of additional nodes of settlement on waterways and inland where roads afforded adequate accessibility. (chapter 2). The more important of these nodes were usually dignified by the setting off of their surrounding townships as a new district, with the emergent town as the seat of the Quarter Sessions and the locus of other administrative functions. Because of the way in which such functions tended to act as foci of attraction for many other kinds of urban activities, there was a good deal of political manoeuvring during the early nineteenth century among local elites to obtain this prized designation. In all, the number of districts increased in this way to a maximum of twenty, attained by 1840. Although this was, in one sense, a continuation of the policy of decentralization, it was not true decentralization, because no further delegation of authority, either of decision-making or of executive functions, took place; rather, the one level of governmental units was fragmented into more and more divisions of the same level. In the same way, the creation of more townships accompanied – indeed, preceded – the spread of settlement inland from the initial areas along shores and waterways. The functions of the early counties continued generally as before, as a special set of subfunctions of districts, although it became common for a county seat (not already a district seat) to have a jail and courthouse facilities and so be an outlier of the district seat for the functions of the Quarter Sessions and higher courts. By 1840, many districts in fact contained but one county, completely conterminous with it, as the earlier large districts containing several counties had been partitioned.

It became evident very early that urban places required some special treatment, since a chronic shortage of qualified magistrates made it difficult for them to deal with the numerous cases arising from the commercial functions of the towns and the greater social interaction that accompanies urban densities. A special category of bodies was devised to cope with this problem, while leaving all the necessary over-riding powers with the magistrates. This was the formation of 'Boards of Police' for the purpose of dealing with the numerous but relatively minor issues specific to the urban community. The first of these was enacted in 1816 (Kingston), followed by other towns in rapid succession. In 1834, the special case of the largest city was recognized by awarding it a charter for local government as a 'self-governing' city founded as the settlement of York in 1793 with the designation of Provincial Capital. The nodality derived from its primate position in the political system was reflected in its rapid growth and increasing complexity and sophistication of its functions. Its elevation to the status of full city government under the name Toronto was, however, essentially an *ad hoc* solution to the problem of urban government, in the manner of the old city charters of Europe; a systematic solution to such problems was, as yet, some fifteen years away, but the Toronto issue foreshadowed the general rise of urban communities to levels which were beyond the capacity of the Quarter Sessions system to handle adequately. More town charters followed, but these tended to be essentially further expansions of the Board of Police concept, rather than full city government. Small urban places were catered for as a group in 1847, by a law providing for limited corporate status for communities with thirty houses or more.

While the process of 'decentralization' of government in the above manner was continuing, there began to appear a need for a partial reversal of the trend at higher levels. With increasing population and commodity movement, there arose pressing needs for social overhead capital in the form of roads, canals, and other fixed investment at a rate greater than the capacity of the local entrepreneurs to provide, in terms of their own personal wealth and the credit they could obtain in money markets. The revenues of government of the time were scanty, the credit rating of the province was poor in consequence, and the financial position of the public treasury was grim. Among other causes, the inability of the provincial government to provide the transport and other facilities seen as essential for its economic development stood out as a rationale for the re-unification of Upper and Lower Canada, effected in 1841. This re-centralization move was in its way a forerunner of homologous tendencies within the respective provinces in the next century or so, and points up an important fact. A society in which the lifestyle becomes more and more complex requires an increasingly larger scale of government activity, in matters of finance, of policing, and of social services (such as education); the capacity of a governmental unit to meet the responsibilities placed on it by its citizens may be overstrained if it is too small in terms of territory and/or population, or if its financial base is inadequate. One result of the Union in 1841 was, indeed, a burst of developmental activity in the construction of canals, the improvement of roads and bridges and harbour facilities, much of it applied to areas somewhat remote from the provincial capital, itself already fairly well served by facilities provided by entrepreneurs.

From the beginning of the decentralization process in 1788, the principal element of local government had been the district, and the whole flavour of government was authoritarian, with decentralization more a matter of the effective implementation of policies decided at higher levels than of the provision for local communities' participation in decision-making. At the same time, the desire of the local communities for decision-making powers which manifested itself originally in the Town Meeting continued to grow, in the context not only of the diffusion of republican ideology from the United States but of changing notions of democracy that were altering the political system of Britain during the earlier nineteenth century. These principles were brought to Upper Canada by immigrants who became politically active, publishing newspapers that expounded these ideals and forming a political party, the Reform Party, that embodied their views of democratic institutions (Craig 1963). These people were primarily interested in decentralization of real decision-making in the Imperial context, which they called 'Responsible Government,' and this concept was extended to the local level as well. The Reformers, then, were decentralizers, although the party included a number of interest groups with overlapping but not identical objectives. In opposition to them were the conservatives ('Tories'), who steadfastly maintained the older view of decentralization as essentially an executive function, and maintained their control of the apparatus of government while the population became more and more alienated from them. One result of this stress was overt rebellion in 1837–8, carried out by extremists in both the upper and the lower province, but put down because there was little active support from the bulk of the population, disenchanted with the Tories though they may have been.

The union of the Canadas in 1841 thus resulted from a complex of events and pressures, but the need for some centralization was among the most cogent. At the same time, the Reformers' philosophy included a notion of further decentralization by making the district councils more representative; from 1842, the councils were composed of delegates elected by each township, and presided over by a warden who was, until 1846, the appointee of the provincial government (Aitchison 1949). At one and the same time, therefore, the two polar tendencies were operating, that of centralization at the level of the province, and that of decentralization at the level of the local community.

Although changes were effected in the functions of government in the early 1840s, there was little if any change in the spatial pattern, except for some additional townships created in zones where new land was being settled. The great changes in the system were brought about in 1849 and the following years by the introduction of the Municipal Corporations Act, commonly known as the Baldwin Act after its originator, and ancillary legislation. In these enactments, the province carried to the maximum extent the principle of functional decentralization by allocating the most important local government functions to the townships, which were made representative through elections by qualified voters of the township councils, which had a great deal of decision-making power. The old districts were abolished; their place was taken by a new type of government at the county level, with restricted duties and powers, and composed of representatives from the township, village, and town councils, but not from city councils (Aitchison 1949). Figure 6.1 shows the spatial extent of township-level government by 1849; the spatial decentralization is very considerable.

Under the Baldwin scheme, separate provision was made for the urban places that had developed, or were expected to develop in the subsequent years. A four-level structure was devised which was related to the rural (township–county) units in the following way. Cities (15,000 population and over) were quite separate for purposes of local self-government, analogous in this respect to counties. Towns (over 3000 inhabitants) and villages (over 1000 inhabitants) were subordinate to county councils in which they were represented by one or two delegates. The lowest level of urban organization was the 'Police Village,' which had no stated minimum population, but was administered as part of a township and had no statutory representatives on either the township or the county council. Throughout the system, residual powers rested with the lowest distinct units, townships, villages, and towns, and not with the counties.

In the abolition of the old districts, and the creation of the new counties, many territorial units did not change, since the expansion of the district system had resulted in a number of instances of a district and a single subordinate county occupying the same area. In these cases, the district was simply dissolved and the county emerged in the same area and with the new functions. Many districts, however, contained several counties. These were commonly erected into the new county units, after they had achieved the minimum population of 15,000 and erected a jail and courthouse to house the officials and provide for the functioning of the local government and judicial system. Some completely new counties were created, especially around nodes of population within older counties that were able to assert sufficient influence to be recognized in this respect. In this way, the old districts, numbering twenty, were replaced by a new set of counties, numbering nearly twice as many. Most of them represented, at least crudely, the areas of influence of established or growing towns and cities.

These county units were probably derived as much as anything from a concept of community-of-interest based on the travel time of foot or horse transport. This parameter was described in 1798 as 'an easy day's journey,' and probably had not changed quantitatively to any great extent by 1850 in the remote districts which still lacked the good roads common in the urbanizing zones on and near the lower lakes. Such units as Grey and Bruce Counties were set up in anticipation of settlement, as were later units in the Canadian Shield in the late nineteenth and early twentieth centuries. In any case, as the extension of the settled area proceeded, the system of local government initiated by the Baldwin Act accompanied it (Fig. 6.2).

There were two interesting exceptions to the Baldwin pattern, however, that reflect the realities of settlement in land areas unfavourable to the development of integrated regional systems of farm and village communities. The county system, and especially its threshold size of 15,000 within any reasonable area proved to be unattainable in Shield conditions. As a result quasi-counties, termed districts (in a quite different political context from the old use of the term) were formed later in the nineteenth century to provide for administration and services in the spaces between the few sporadically distributed settled townships that could attain municipal status. Second, urban nucleations of small size, which would qualify only as a village in the closely settled parts of the province, would have the powers and structure of a town; by this means the settlement was entitled to its own magistrate, and so would not have to depend on the visits of travelling magistrates or incur the expense of sending police prisoners to a distant place for hearings.

This pattern was continued northwards beyond the limits of Ontario as they were in 1867, into territory added to the province from 1874 to 1912 from the former Hudson's Bay Company lands. In all of the districts of

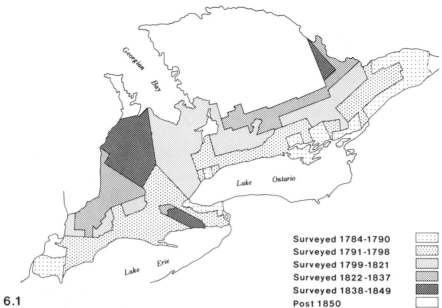

6.1

Development of the Township Pattern

Surveyed 1784-1790
Surveyed 1791-1798
Surveyed 1799-1821
Surveyed 1822-1837
Surveyed 1838-1849
Post 1850

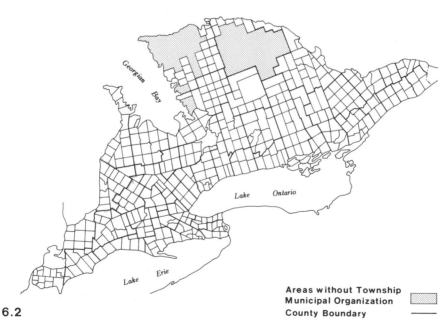

6.2

Municipal Townships in 1966

Areas without Township
Municipal Organization
County Boundary

the Canadian Shield, farming was sporadic and marginal; the urban places that have arisen mostly derive from mining activities or transportation nodes formed by the infrequent junctions of a very open web of railways (see chapter 2). Thus there is a large number of 'towns' in the north of Ontario that would not qualify for this legal status in the south.

It is ironic, perhaps, that as the greatest degree of political decentralization was being achieved in the 1850s, the tide was already turning towards economic centralization, in response to the growth of urbanism and the requirements of an increasingly complex economic system. This was foreshadowed, as noted earlier, by the incorporation of Toronto and by the union of the Canadas at a higher level. Indeed, if the union of the Canadas was the context in which Baldwin and his Reform colleagues succeeded in accomplishing the ideal in local government (in their terms), it also provided the basis for the changes that rendered their ideal obsolete. In the same year that the Baldwin Act was passed (1849) another act (the Railway Guarantee Act, 1849) stimulated the construction of railways, delayed previously by the problems of raising capital for long-distance routes. These railways almost everywhere generated much urban growth and fostered the development of factory industries. The great bulk of the later growth occurred in the cities, towns, and large villages that were in existence in 1849 (see chapter 2, Fig. 2.3), and which were linked together by the earliest railways of the mid-nineteenth century. The expansion of railways into other parts followed, mostly in the 1870s and 1880s, and the immediate impact on land settlement was striking; but the communications that drew people onto the land in one period were later the means for them to leave. Better economic opportunities were perceived in the growing towns and, for those with a preference for farming, on newer frontiers opening in the interior plains regions of the United States and Canada. Many rural units attained their maximum population in the period 1881–1901 (Watson 1947). They have since been declining in absolute numbers, and even more in their relative position, with increasing heartland and hinterland contrasts as a result (see chapter 3). This is one of the problems faced by present policy makers in Ontario.

The fixed size of the township, the fact that its revenues were very largely from rates levied against real property, and the fact that many townships were undergoing chronic out-migration in the last quarter of the nineteenth century constituted a built-in rigidity that was (and is) ill-suited to cope with the increasing demands of the urbanizing community of the twentieth century. As an example of this, it may be noted that beginning in the 1880s many special functions associated with the changing life style of the province were taken from the townships and given to spe-

City
Town
Village

Lake Ontario

Lake Erie

Metropolitan Toronto

50 Miles
80 Kilometers

200 Miles
320 Kilometers

6.3

Urban Municipalities, 1966

cial boards. Public Libraries, Parks, Health, Sanitation, Welfare, and Public Utilities are all semi-autonomous boards and commissions, over which the elected councils have little or no statutory control.

Because supervision of these bodies, and a considerable portion of their finances, come from the Province of Ontario, they can be considered to represent elements of the trend towards centralization or, at least, recentralization. Still further control over the municipalities was provided through the creation of the Ontario Municipal Board in 1932, and later of the Department of Municipal Affairs, especially under the stress of the Depression in the 1930s. This board has veto power over land use and development schemes.

The problem of the expansion of the area of growing cities since 1849 has been met until recently by simple annexation of land to the city corporation from surrounding townships. As a result, their areas, population, and financial resources were reduced but not their administrative structure. For over a half-century, this process worked well enough, but with the arrival of the motor car, the spatial patterns of Ontario, as elsewhere, underwent a drastic change. The influence of towns and cities extended now for many miles into the surrounding rural areas, rather than the very few miles that was normal up to 1914. The emergence of new communities of interest far transcending the territories of a single city and its adjacent townships, and the effective blurring of the older distinction between the life styles of city and country, all point towards the need for new forms of local government to deal with the new aspects of society (Fig. 6.3). Nowhere is the problem more evident than in the Shield areas, where city dwellers (many from outside Ontario) have purchased land on lakes and rivers for summer cottages and even year-round recreational use (see chapter 3).

These non-residents now greatly outnumber the permanent residents in areas where only a handful of farmers, loggers, or transport workers had comprised the township electorates of previous years. Now the city dwellers are demanding (and obtaining) representation on these councils, to effect improvements in the services the township provides, to bring about land-use planning and other aspects of a more urban life style for which the original residents in the townships had little need and less desire.

Coupled with the problems of economic decline, both relative and absolute, in many rural areas, the manifest obsolescence of the Baldwin system has been the subject of a number of official studies, and several results of these studies have been implemented to date. Perhaps the first of the recent attempts to effect some recentralization in Ontario occurred with the formation in 1954 of a supercity organization, the Municipality of Metropolitan Toronto, introduced to provide essential co-ordination among a number of municipal units which had become functionally linked in the expansion of Toronto (see chapter 5), namely the City of Toronto, five highly urbanized townships, and seven towns and villages that had, in the Baldwin tradition, grown up around minor nodes in the vicinity of the city.

Another aspect of recentralization is exemplified by the grouping of municipalities, rural and urban, in Authorities for regulation of water use and the encouragement of conservation. The forerunner of this practice was the Grand River Commission, formed in 1940 primarily to construct flood-mitigation works. After World War II, however, in the context of the reconstruction of the economy, a wider arrangement was introduced in the Conservation Authorities Act, in which a voluntary pooling of powers of local municipalities within a watershed is permitted for common uses of the public water within the area.

The policy of the Ontario government is bringing about a new form of regional municipality, better attuned to the needs of the twentieth century, in the economic and political sense (Ontario Department of Municipal Affairs 1966). It is recognized that a region should exhibit a sense of community; a region should have a balance of interests; there must be an adequate financial base; the region should be large enough so that local responsibilities can be performed efficiently; and regional boundaries should make possible maximum co-operation between regions.

The implementation of the policy has, to the end of 1970, taken the form of special studies and legislation for specific areas, in the main those in which the problems of rural–urban stress and urban disfunctioning were most severe. Commissions to study some fourteen areas were set up during the 1960s, and a number of these have resulted in new units of local government, generally within the framework of the policies outlined above. The earlier version of Metropolitan Toronto was restructured in 1966 into the city and five boroughs, supplanting the other twelve municipal units of the 1954 plan. Regional municipalities were formed out of the city of Ottawa and Carleton County; of the two counties of Lincoln and Welland and all urban and rural units within their territorial limits; of the balance of York County outside Metro Toronto; of the resort district of Muskoka; and of the urban complex at the head of Lake Superior (now named the Regional Municipality of Thunder Bay). More may be forthcoming.

As has been graphically shown in the earlier chapters of this monograph, the economic pattern of Ontario has become strongly concentrated into a relatively few actively growing centres, with the province's heartland of the Golden Horseshoe (the western littoral of Lake Ontario) attracting by far the largest share of economic activity and migration (Fig. 6.3). The resulting heartland–hinterland contrasts have caused concern that this rapid economic centralization has not been altogether healthy, socially and practically, for the people of Ontario. The remedies that are currently being discussed and applied are chiefly those of economic incentives to individual firms. But there is a strong possibility that the issue of territorial re-organization of the province in the political and administrative sense may be closely articulated with management of the spatial–economic system. The political recentralization required for the better functioning of local government will thus meet (at a 'regional' level) the economic decentralization increasingly perceived as a desirable social goal. If this indeed does happen, the fit between local government area and spatial economic unit may be harmonious again, for a time. Such new sets of political–territorial units could be a functional framework for any attempt to redress the heartland–hinterland disparities that have been so long in the making.

References

Aitchison, J.H., 1949   The Municipal Corporations Act of 1849, *Canadian Historical Review*, 30, 2 (June): 107–22

Alexander, J.W., 1954   The Basic-Nonbasic Concept of Urban Economic Functions, *Economic Geography*, xxx, 3 (July): 246–61

Alexandersson, G., 1956   *The Industrial Structure of American Cities* (Lincoln, Nebraska)

Arthur, Eric, 1965   *Toronto: No Mean City* (Toronto)

Berry, Brian J.L., 1965   Identification of Declining Areas: An Empirical Study of the Dimensions of Rural Poverty, *in* W.D. Wood and R.S. Thoman (eds.), *Areas of Economic Stress in Canada* (Kingston): 22–66

— 1967   *Geography of Market Centers and Retail Distribution* (Englewood Cliffs, New Jersey)

Boulding, K.E., 1953   Toward a General Theory of Growth, *Canadian Journal of Economics and Political Science*, xix, 3 (Aug.): 337

Bourne, Larry S., 1967   *Private Redevelopment of the Central City: Spatial Processes of Structural Change in the City of Toronto* (Chicago)

— 1968   Market, Location and Site Selection in Apartment Construction, *Canadian Geographer*, xii, 4 (Winter): 211–26

Burghardt, A.F., 1969   The Origin and Development of the Road Network of the Niagara Peninsula, Ontario, 1770–1851, *Annals, Association of American Geographers*, 59, 3 (September): 417–40

Canada, Dominion Bureau of Statistics, 1960   *Standard Industrial Classification Manual*

— 1961   *Census of Canada* (Ottawa: Queen's Printer)

— 1966   *Census of Canada* (Ottawa: Queen's Printer)

— 1969   *Growth Patterns in Manufacturing Employment by Counties and Census Divisions* (Ottawa: Queen's Printer)

Canada Land Inventory, 1966   The Climates of Canada for Agriculture (Ottawa: Canada Department of Forestry and Rural Development)

— 1970   *Land Use Capability for Agriculture* (Toronto: A.R.D.A. Branch, Ontario Department of Agriculture and Food)

Carol, Hans, 1969   Development Regions in Southern Ontario Based on City-Centred Regions, *Ontario Geography*, 4: 13–29

Cass-Beggs, D., 1970   Water as a Basic Resource, *in* R.R. Krueger et al. (eds.), *Regional and Resource Planning in Canada* (Toronto): 176–97

Centre for Urban and Community Studies *Research Papers*   A series of working papers published by the University of Toronto, several of which deal specifically with growth and change in Toronto:

ON LAND USE

No. 25, Günter Gad (1970), A Review of Methodological Problems in Estimating Urban Expansion
No. 31, L.S. Bourne (1970), Dimensions of Metropolitan Land Use: Cross-Sectional Structure and Stability
No. 35, M.J. Doucet (1970), Trends in Metropolitan Land Use and Land Consumption: Metropolitan Toronto

No. 38, L.S. Bourne and M.J. Doucet (1970), Dimensions of Metropolitan Physical Growth: Land Use Change, Metropolitan Toronto

ON CHANGE IN THE TORONTO URBAN FIELD

No. 28, G. Hodge (1970), Patterns and Parameters of Industrial Location in the Toronto Urban Field

No. 29, G. Hodge (1970), Cottaging in the Urban Field: A Probe of Structure and Behavior

No. 30, G. Hodge (1970), A Probe of Living Areas in the Periphery of the Toronto Urban Field

ON POPULATION DENSITIES

No. 34, F.I. Hill (1970), Spatio-Temporal Trends in Population Density: Toronto

ON REGIONAL INTERACTION

No. 44, J.W. Simmons (1971), Net Migration within Metropolitan Toronto

Chapman, L.J. and D.F. Putnam, 1966   The Physiography of Southern Ontario (Toronto)

Chatelain, A., 1957   Géographie sociologique de la presse et régions françaises, Revue de Géographie de Lyon, XXXII: 127–34

Christaller, W., 1933   Die zentralen Orte in Süddeutschland, translated by C.W. Baskin as Central Places in Southern Germany (Englewood Cliffs, New Jersey, 1966)

Coyne, J.H., 1908   The Talbot Papers (Ottawa): 37

Craig, G.M., 1963   Upper Canada 1784–1841 (Toronto)

Cruickshank, E.A. (ed.), 1923   The Simcoe Papers (Toronto), I: 120

Dean, W.G. (ed.), 1969   Economic Atlas of Ontario (Toronto)

Doxiadis Associates, 1969   The Developing Great Lakes Megalopolis Research Project Inc.: Summary and Evaluation of the Studies Conducted within the UDA Research Project for the Great Lakes Megalopolis (Detroit)

Duncan, O.D. et al., 1960   Metropolis and Region (Baltimore)

Economic Council of Canada, 1965   Toward Sustained and Balanced Economic Growth: Second Annual Review (Ottawa: Queen's Printer)

Firth, E.G., 1962, 1966   The Town of York, I, 1795–1815; II, 1815–1834 (Toronto)

Friedmann, John et al., 1970   Urbanization and National Development: A Comparative Analysis (MS thesis, School of Architecture and Planning, University of California at Los Angeles): 5

Gentilcore, R.L., 1963   The Beginnings of Settlement in the Niagara Peninsula, The Canadian Geographer, VII, 2: 72–82

— 1969   Lines on the Land, Ontario History, LXI, 2 (June): 57–73

Gertler, Leonard O., 1969   A Concept for Delimiting Development Regions, Ontario Geography, 4: 30–4

Goheen, Peter G., 1970   Victorian Toronto, 1850 to 1900: Pattern and Process of Growth (Chicago)

Guillet, E.C., 1957   The Valley of the Trent (Toronto): 84–130

Hamelin, Louis-Edmond, 1968   Types of Canadian Ecumene, *in* Robert M. Irving (ed.), *Readings in Canadian Geography* (Toronto): 20–30

Hills, G.A., 1961   The Ecological Basis for Land Use Planning (Toronto: Ontario Department of Lands and Forests)

Innis, H.A., 1967   The Importance of Staple Products, *in* W.T. Easterbrook and M.H. Watkins (eds.), *Approaches to Canadian Economic History* (Toronto): 19

Jones, R.L., 1946   *History of Agriculture in Ontario* (Toronto)

Kaplan, Harold, 1967   *Urban Political Systems: A Functional Analysis of Metropolitan Toronto* (New York)

Kasahara, Yoshiko, 1963   A Profile of Canada's Metropolitan Centres, *Queen's Quarterly*, 70, 3 (Fall): 311

Kerr, Donald P., 1968   Metropolitan Dominance in Canada, *in* John Warkentin (ed.), *Canada: A Geographical Interpretation* (Toronto): 537–43

Kerr, D. and J. Spelt, 1965   *The Changing Face of Toronto – A Study in Urban Geography* (Ottawa: Geographical Branch Memoir 11, Department of Mines & Technical Surveys)

King, Leslie J., 1966   Cross-Sectional Analysis of Canada Urban Dimensions 1951 and 1961, *Canadian Geographer*, x, 4 (Winter): 205–24

Kirk, D.W., 1949   The Areal Pattern of Urban Settlements in Southwestern Ontario in 1850 (PHD thesis, Northwestern University)

Koenig, E. Frederick, 1971   Heartlands, Hinterlands and Allometric Growth (MA thesis, Department of Geography, State University of New York at Buffalo)

Konarek, J., 1970   Algoma Central and Hudson Bay Railway: The Beginnings, *Ontario History*, LXII, 2 (June): 75

Lithwick, N. Harvey, 1971   *Urban Canada: Problems and Prospects* (Ottawa: Central Mortgage and Housing Corporation)

Lorimer, James, 1970   *The Real World of City Politics* (Toronto)

Mackay, J. Ross, 1958   The Interactance Hypothesis and Boundaries in Canada, *Canadian Geographer*, 11 (Spring): 1–8

MacKinnon, Ross D. and John Hodgson, 1969   The Highway System of Ontario and Quebec, *Research Paper No. 18* (Toronto: University of Toronto, Centre for Urban and Community Studies)

Marshall, J.U., 1969   *The Location of Service Towns: An Approach to the Analysis of Central Place Systems* (Toronto)

Matthews, B.C., 1956   Soil Resources and Land Use Hazards in Southern Ontario, *Canadian Geographer*, 8: 55–62

Maxwell, J.W., 1965   The Functional Structure of Canadian Cities: A Classification of Cities, *Geographical Bulletin*, VII, 2: 79–104

Metropolitan Toronto and Region Transportation Study (MTARTS), 1967   *Choices for a Growing Region* (Toronto)

Miller, M.H. and D.W. Hoffman, 1970   The Challenge of the Land, Physical Resource Aspects, *in* D. Elrick, *Environmental Change, Focus on Ontario* (Toronto): 9–30

Murdie, Robert A., 1969   *Factorial Ecology of Metropolitan Toronto, 1951–1961* (Chicago)

Naroll, Raoul S. and Ludwig von Bertalanffy, 1956   The Principles of Allometry in Biology and the Social Sciences, *General Systems*, vi: 76–89

Nelson, H.J., 1955   A Service Classification of American Cities, *Economic Geography*, xxxi, 3 (July): 189–210

Noble, Henry F., 1965   *Variation in Farm Income of Farms in Eastern Ontario by Farm Type and Farm Class and An Economic Classification of Farms in Eastern Ontario* (Toronto: Ontario Department of Agriculture and Food)

Ontario Department of Municipal Affairs, 1966   *Design for Development: Statement of the Prime Minister on Regional Development Policy* (mimeographed)

Ontario Economic Council, 1970   *Trends, Issues and Possibilities for Urban Development in Central and Southern Ontario* (Toronto)

Ontario. Regional Development Branch, 1968   *The Niagara Escarpment Study* (Toronto: Department of Treasury and Economics)

Ontario. Regional Development Branch, 1970   *Design for Development: The Toronto-Centred Region* (Toronto: Department of Treasury and Economics)

Ontario. Vital Statistics Report   Annual (Toronto: Department of the Registrar General)

Pratt, R.T., 1968   An Appraisal of the Minimum-Requirements Technique, *Economic Geography*, xliv, 2 (April): 117–24

Pred, A., 1965   Industrialization, Initial Advantage, and American Metropolitan Growth, *Geographical Review*, lv, 2 (April): 158–85

Preston, R.A., 1959   *Kingston Before the War of 1812* (Toronto)

Ray, D. Michael, 1965   *Market Potential and Economic Shadow* (Chicago): 89–110

— 1969   The Spatial Structure of Economic and Cultural Differences: A Factorial Ecology of Canada, *Papers of the Regional Science Association*, xxiii: 7–23

Reeds, L.G., 1956   Agricultural Geography of Southern Ontario (PHD thesis, University of Toronto): 24

Richmond, Anthony H., 1967   Immigrants and Ethnic Groups in Metropolitan Toronto, *Research Report E-1* (Toronto: York University, Institute of Behavioural Research)

Robinson, J. Lewis, 1969   *Resources of the Canadian Shield* (Toronto): 3

Rose, A.J., 1966   Dissent from Down Under: Metropolitan Primacy as the Normal State, *Pacific Viewpoint*, vii, 1 (May): 1–27

Schott, C., 1936   *Landnahme und Kolonisation in Canada, Am Beispiel Südontarios* (Kiel): 212–21

Shortt, A., 1902   The Beginning of Municipal Government in Ontario, *Canadian History*, v, 7: 409–24

Simmons, James W., 1968   Changing Residence in the City: A Review of Intraurban Mobility, *Geographical Review*, lviii, 4 (October): 621–51

— 1970   Patterns of Interaction within Ontario and Quebec, *Research Paper No. 41* (Toronto: University of Toronto, Centre for Urban and Community Studies)

Smailes, A.E., 1966   *The Geography of Towns* (London)

Smallwood, Frank, 1963   *Metro Toronto: A Decade Later* (Toronto)

Smith, R.H.T., 1965   The Functions of Australian Towns, *Tijdschrift voor Economische en Sociale Geografie*, LVI, 3 (May/June): 81–92

Spelt J., 1955   *The Urban Development in South-Central Ontario* (Assen). Reprinted in Carleton Library Series (Toronto, 1972)

— 1968   Southern Ontario, *in* John Warkentin (ed.), *Canada: A Geographical Interpretation* (Toronto): 348–65

Stone, Leroy O., 1968   Urban Development in Canada (Ottawa: Census Monograph, Dominion Bureau of Statistics)

Thompson, W.R., 1965   *A Preface to Urban Economics* (Baltimore)

Ullman, E.L. and M.F. Dacey, 1960   The Minimum Requirements Approach to the Urban Economic Base, *Papers and Proceedings of the Regional Science Association*, VI: 175–94

Warkentin, J., 1966   Southern Ontario: A View from the West, *Canadian Geographer*, X, 3: 170

Watson, J.W., 1947   Rural Depopulation in Southern Ontario, *Annals of the Association of American Geographers*, 37: 145–54

Webb, J.W., 1963   The Natural and Migrational Components of Population Changes in England and Wales, 1921–1931, *Economic Geography*, XXXIX, 2 (April): 130–48

Whebell, C.F.J., 1968   Net Migration Patterns 1956–61 in Southern Ontario, *Ontario Geography*, II: 67–81

— 1969   Corridors: A Theory of Urban Systems, *Annals of the Association of American Geographers*, 59 (March): 1–26

Wolfe, Roy I., 1968   Economic Development, *in* John Warkentin (ed.), *Canada: A Geographical Interpretation* (Toronto): 226–27